Gal Luft and Anne Korin

TURNING OIL INTO SALT

Energy Independence Through Fuel Choice

To our parents, who gave us both energy and security

Technology is a real enemy for OPEC.
Sheikh Ahmed Zaki Yamani, former Saudi oil minister

CONTENTS

INTRODUCTION:
FROM THE BEACH TO THE BASEBALL DIAMOND

One thousand dollars a year. That was the salary of America's first oilman, Colonel Edwin Laurentine Drake, hired by the Seneca Oil Company to look for oil in Titusville, Pennsylvania. After long months of aggravation came success and in August 1859, 150 years prior to the publishing of this book, the first barrel was discovered and lifted from a depth of 69.5 feet, giving rise to America's most profitable industry. Despite occasional wars fought over oil, during its century-and-a-half long history the black gold has brought the world much good. Petroleum has enabled the production of industrial chemicals, heating oil, medicines, plastics, asphalt and lubricants, all of them critical to our modern society. But most importantly, it has powered mobility. Today, roughly two-thirds of the world's oil is used for transportation, and most vehicles are able to run on *nothing* but it. And that is exactly the problem. As author Robert Zubrin once observed, *we* are not "addicted to oil." Our cars are.

This book argues that the threat oil dependence presents to our national and economic security is not a function of the amount of oil we consume or import. It is a function of oil's status as a strategic commodity. Oil's strategic status stems from its virtual monopoly over fuel for transportation, which underlies the global economy and our entire way of life. Without oil, food cannot travel from farm to plate, mail cannot reach its destination, raw materials cannot reach their factories and children cannot attend their schools. Indeed, without oil, every child is left behind. Oil is the mortar that holds our economy together.

To understand the implications of overdependence on a strategic commodity and what can be done to diminish the destructive status of certain strategic commodities, we can look to history. Years ago, salt was a strategic commodity, one that determined the course of world affairs and the stature of countries on the global stage. Wars were fought over salt, and colonies were formed in remote places where it happened to be found. That was because salt had a virtual monopoly over food preservation. With the advent of canning, electricity, and refrigeration, salt lost its strategic status. We still use and trade salt – the worldwide average for salt intake per individual is about 10 grams per day - but we would bet most readers don't know who the world's big salt reserve holders are, and whether or how much salt the United States imports. And we would bet regardless of the data, most would not say that the United States has a "salt dependence" problem (though certainly some of us as individuals do.) We are "salt independent" regardless of the amount of salt we consume or import. We are "salt independent," because salt is just not that important any more. No American president would even think of bowing to the world's largest salt reserve holder as President Obama did before Saudi Arabia's King Abdullah in their April 2009 meeting.

This is what energy independence means: that it no longer matters who holds the reserves, that oil becomes much less relevant to global affairs, that it becomes just another commodity. Contrary to popular conception, energy independence does not mean autarky - it doesn't mean walling ourselves off from the global market. Independence means not having to kowtow to the various petrodictators that sit on the bulk of the world's oil reserves. Independence requires that oil become just not that important any more.

Due to rapid economic growth and declining domestic reserves, less than a century after Titusville, the United States became a net oil importer, and its dependence on foreign oil has increased by leaps and bounds ever since. As has become abundantly clear from the events of recent years, this dependence gives a small group of nations, spearheaded by members of the Organization of Petroleum Exporting Countries (OPEC) and Russia, control over the global transportation system and by extension, over the world economy at large. The combination of oil's monopoly over transportation fuels and a manipulative cartel's control over most of the world's oil reserves is paralyzing our foreign policy, bankrupting our economy and at the same time bankrolling our enemies. Because the American way of life is one of the most oil dependent in the world – most Americans cannot even buy their groceries or see a doctor without driving somewhere – this dependence is "the albatross of U.S. national security" as Indiana Senator Richard Lugar once put it.[1] The rise of Islamic fundamentalism, the never ending instability in the Middle East and Africa, terrorism and piracy against oil facilities and tankers and the natural disasters in energy producing areas are constant reminders that our oil supply, and, by extension, our way of life, is increasingly vulnerable. And yet, every year, we continue to put on our roads in the United States alone more than 10 million new petroleum-only cars, each with a street life of over 15 years, hence locking our future to petroleum exporting nations and their whims for many years to come.

A manipulative cartel

While members of OPEC sit on some 78 percent of world oil reserves, they account for only 40 percent of global oil production due to a deliberate strategy of constraining supply. In 1973, just before the Arab Oil Embargo, OPEC produced 30 million barrels per day. Thirty-six years later, with global oil demand and non-OPEC production nearly doubling – and despite the fact that in 2007 the cartel inducted two new members, Angola and Ecuador, with combined daily production capacity equivalent to that of Norway – OPEC's crude production has not increased. In fact, in response the economic downturn of 2008-09 – driven in part by the loss of disposable income caused by high energy prices – OPEC throttled supply down, and in 2009 it is expected to average 29 million barrels a day – one million barrels per day less than in 1973! OPEC's manipulation of oil prices by fluctuating supply is designed for one end: maintaining the virtual monopoly of oil over the global transportation sector while blocking competition from alternative energy sources. True to form, in response to President Barack Obama's call to steer the country's energy policy away from oil, the Saudi Oil Minister Ali Naimi, the man who effectively runs OPEC, resorted in 2009 to the old scare tactics to dissuade oil importing nations from promoting alternatives. He warned that efforts to rapidly promote alternatives could have a "chilling effect" on investment in the oil sector and that "a nightmare scenario would be created if alternative energy supplies fail to meet overly optimistic expectations, while traditional energy suppliers scale back investment."[2] Translation: if you keep investing in alternative fuels, we are going to cut back on investment to make sure oil prices stay high.

March of fallacies

The combination of oil's monopoly in the transportation sector and OPEC's control over world reserves is toxic. Failure to address it today will inevitably lead to a situation described by the International Energy Agency's chief economist: "we are ending up with 95 percent of the world relying for its economic well-being on decisions made by five or six countries in the Middle East."[3] Over the years, we have worked to prevent such a dim future, appearing before hundreds of domestic and international forums throughout the world and briefing numerous members of Congress, foreign leaders, business executives, governors, religious leaders, pundits, war veterans and even presidential candidates. While making the case for a shift from oil has become an increasingly easy one to make, we have often been struck by the degree of ignorance and lack of clarity on the issue when it has come to solutions. A public opinion poll published in April 2009 reveals the depth of the problem: forty percent of those surveyed couldn't name one fossil fuel, and 50 percent couldn't even name one source of renewable energy.[4] Our elected officials also have much to learn. Take the 2008 presidential campaign for example. Unlike in the Seventies, when a significant portion of our electricity was generated from oil, today only 2 percent of U.S. electricity is oil powered, and conversely only 2 percent of our oil demand is due to electricity generation. This basic fact did not prevent Republican candidate Senator John McCain from touting nuclear power as the key to ending our oil dependence while his Democratic opponent, Senator Barack Obama, highlighted solar and wind power. There is nothing wrong with these electricity sources. However, shuffling around what we generate electricity from is a separate issue from reducing oil dependence since we essentially no longer generate electricity from oil. Build as many solar panels or nuclear reactors you wish and you might displace coal or natural gas – but not oil. Plug in an electric car today and it will for the most part run on something other than oil. You'd expect those who run for the highest office to know their Energy 101.

The march of fallacies gets worse. Some people, for example, propose that buying less oil from the Middle East and more from friendly nations would improve our position. Sounds so reasonable. But oil is a fungible commodity. Think of the oil market as a swimming pool: producers pour oil in, consumers take oil out. We don't import all or even most of our oil from the Middle East. In fact, the Middle East is source of barely one quarter of our imports. But that doesn't lessen the control Middle Eastern countries, which sit on some two thirds of world oil reserves, have over the world market. We don't import a drop of oil from Iran, but anything that impacts Iranian supply affects the whole market, affects oil prices for everyone, not just those who buy directly from Iran. If Iran's President decided to cut the country's oil exports or block the Straits of Hormuz, the increase in price would be felt across the board. Therefore, if we just shuffle around our sources of oil supply by buying oil from different countries than we do today, it will not reduce our vulnerability to the oil cartel's market manipulation. Someone else will buy the oil we would have bought from supplier A, and we will buy the oil they were buying from supplier B – no difference whatsoever will be felt by the suppliers (or us).

Further, despite the fact that much of the public debate on this issue has been dominated by calls to "drill more" (Republicans) or to "use less" (Democrats), the fact is that drilling and efficiency are two sides of the same coin. They both perpetuate the petroleum monopoly in transportation, doing nothing to weaken OPEC. As we show in Chapter 3, when non-OPEC countries drill more, OPEC simply drills less, and when we use less, OPEC also drills less – witness the multiple production cuts in 2008 and 2009 geared to propping up oil prices. Neither efficiency nor drilling will serve to strip oil of its strategic status. At best, these policies could keep some of our dollars from migrating overseas.

We decided to write this book in an attempt to clear some of these common misconceptions and to reorient the public discourse toward solutions that actually make a difference. We began the work in the summer of 2008 when oil prices where at record level of nearly $150 a barrel. Gasoline prices in the United States neared $5 a gallon, and the U.S. economy bled more than half a trillion dollars in oil imports – more than our defense budget. Oil dependence was the topic du jour. Then came the worst economic crisis in recent history and shuffled the deck. Demand for oil and its products dropped significantly, driving down prices by nearly $100 per barrel. New oil production stalled, and alternative energy projects were shelved. As we write these words, the U.S. auto industry is effectively bankrupt, the financial system is still in disarray and our dependence on foreign oil has been relegated to the back burner of our national priorities.

But not for long. The nature of recessions, deep as they may be, is that they are transitory. Sooner or later, America will zoom out of its economic predicament carrying with it the rest of the world economy. Growth will resume and with it our appetite for oil. Once that happens, we can expect prices to hit record highs again due to OPEC supply constraints in the face of developing world demand growth, and with high prices will continue the wealth transfer from consuming countries to oil exporters like Saudi Arabia and Iran, the former of which funds and propagates radical Sunni Islam and the latter radical Shia Islam. At this point those two countries, as well as others in the Middle East, could well be proud owners of a petrodollar-funded nuclear arsenal.

Geology is also raising its head. To understand what lies ahead take a glimpse into the International Energy Agency (IEA) World Energy Outlook published in late 2008. The IEA report examined the world's 800 top oilfields and reported an average annual depletion rate of 5.1 percent increasing to 8.6 percent in 2030.[5] It stated that in order to meet future demand for oil, four new Saudi Arabias will have to be added to the global oil market between now and 2030. What is the chance of this happening? Even if such a gigantic amount of capacity can ever be added it would require trillions of dollars of investment in new exploration and recovery as well as the development of non-conventional sources of oil such as tar sands and oil shale. But the economic conditions of 2008-2009 have thwarted much needed investment in new production. According to OPEC, during the recession 35 major exploration projects have been shelved. Ali al Naimi, the Saudi oil minister, warned in the March 2009 OPEC meeting of a coming "catastrophic" shortfall in petroleum production, raising doubts the world can count on the one Saudi Arabia that exists, not least the four that don't. This means that once demand for oil returns to its previous level, an oil crisis of epic proportions could ensue, throwing the global economy into yet another economic crisis. Therefore, it would be prudent that as we struggle to pull ourselves out of the current crisis we keep our eyes on the day after and make the necessary preparations for the next era of high oil prices.

Our energy problem is lack of fuel choice

Stripping oil of its strategic status, or if you want, turning oil into salt, begins with the understanding that we must change the game. Playing in the same playing field with the likes of Hugo Chavez, Saudi King Abdullah, Mahmoud Ahmadinejad and Vladimir Putin is playing a game we can never win. They have most of the world's oil; we have – drill everywhere – barely three percent of conventional oil reserves. The sooner we adjust our thinking and focus on game-changing, transformational solutions, instead of inconsequential, time buying policies, the sooner we can reach true and lasting energy independence. In order to strip oil of its strategic status, we must break its monopoly in the transportation sector through fuel choice. Think about it. We have choice in every aspect of our lives except for transportation fuel. Walk into the nearest Starbucks and you can choose among dozens of beverages. At the supermarket a friendly cashier asks you "paper or plastic?" At dinner the waiter asks: "red or white wine?" And in the air, flight attendant offers us "coffee or tea?" Whether we buy a carpet, a pair of glasses, a laptop, a bowl of soup or an insurance policy, choice is everywhere. Not so at the fuel station. Here its gasoline, gasoline or gasoline made from oil, oil and oil. This lack of choice is what lies at the core of our energy security challenge.

During a March 2008 Congressional hearing on oil dependence conducted by the U.S. House of Representatives Committee on Foreign Affairs, Congressman Bob Inglis, a South Carolina Republican, provided perhaps the best metaphor depicting our predicament with regards to oil:

> I wonder if we are sort of like the guy that goes out on the beach, and he thinks he is doing all right, except he is one of these guys like me that turns into a lobster at the beach. Along comes this most unfortunate event when the lifeguard starts pulling down the stand, and he is an incredibly fit, you know, tanned fellow, and we are now feeling like wimps because we have got 3 percent of the world's known oil reserves, and it hurts our feelings. [...] We have got these little bodies. But the guy that turns into a lobster at the beach may be a fabulous baseball player. [...] So you change the game. You get off the beach, and you say, "Come on, fella. Let us go to the baseball diamond, and let us compete there." So rather than go to the baseball diamond, we keep slugging it out on the beach, hoping to put on more sunscreen and get the shirts on [...] so then we could cover up our problem.

This journey from the beach to the baseball diamond is what this book is about.

⌘　⌘　⌘

1
UNDERSTANDING STRATEGIC COMMODITIES

Our modern society, with all its hustle and bustle, is powered by three principal forms of energy. The first is food. Each and every one of us needs thousands of calories every day to live a productive and healthy life. When our food intake drops significantly, lethargy and physical decline soon follow. While food is the most elementary form of energy used by every organism, our second form of energy, electricity, is a feature of the modern era. It heats, cools and lights our homes, powers our computers, refrigerates our food and is essential to our flow of information. When the power goes off, we immediately slide back to the pre-industrial age. Third is transportation energy, the form of energy that moves the world around. Transportation energy is what powers our cars, trucks, ships, planes and trains, enabling the global flow of people, goods and services. Transportation energy also allows the two previous forms of energy to reach their consumers. Our food travels on average 1,500 miles from farm to plate, and raw materials that make our electricity like coal, natural gas and uranium travel thousands of miles across the globe from the mine to the power station.

Throughout history, the delivery of each of the above three forms of energy has either been dominated, or still is, by a strategic commodity without which the energy would not be easily available to its users. For thousands of years, it was salt that humans used to preserve food. From the 19th century, electricity production has been dominated by coal. And the transportation sector since the shift away from horses, carts, coal fire steam ships and trains, is all about oil. Once geology revealed their prevalence, salt, coal and oil became the most sought-after commodities, shaping the borders and international behavior of empires and triggering major world events like wars and revolutions. In other words, they became strategic commodities.

What makes a commodity strategic?

Many of the commodities traded on the world's commodity exchanges from Chicago to Kuala Lumpur are strategic to a particular industry they serve. For the ethanol industry corn is strategic; the clothing industry is highly dependent on cotton; and for the candy industry cocoa and sugar are essential. Aluminum, wheat, rice, iron ore, gold and silver all have unique and essential uses to someone. Are they all strategic commodities in the geopolitical sense? Clearly not. So what makes a commodity *strategic* versus *just another commodity?* A commodity doesn't become strategic just because it is used by most people. Most people drink coffee, and that makes coffee a dominant commodity in the universe of beverages, but that doesn't make coffee a strategic commodity. People may prefer coffee to other beverages perhaps because of its flavor and invigorating properties. But should something happen to our supply of coffee beans, we could start our day with tea or orange juice just as well. We certainly are not going to invade Colombia or Brazil to secure our cup o' Joe.

We can look to history to better understand this matter. Between the fourth and the first millennium B.C., almost every civilization went through a phase in its cultural development which historians call the Bronze Age, the period when metalworking enabled the creation of bronze artifacts. It was not the bronze that was a strategic commodity but the three commodities from which it was made. Bronze was the marriage of copper and tin. Alloyed together, at high temperatures enabled by the third commodity, charcoal, the two naturally occurring ores created a new hard metal that was used to manufacture tools which humanity had never before seen. Bronze axes, swords, sculptures and boats changed the world through peace and war. An entire civilization, the Minoan civilization, rose in Crete, specializing in the tin trade. Those who had tin or controlled its trade routes dominated the world. Those who didn't had to fight for it. But in 1800 B.C., tin became scarce, the Minoan civilization began to collapse, and people were forced to seek alternatives to bronze. Thus began the Iron Age. Iron, much more abundant than its predecessors, became available to many. Its melting point was higher than that of bronze, the tools made from it were stronger, and as production techniques improved, it was less brittle. While iron production required no tin, it still relied on charcoal, which continued to remain a strategic commodity for centuries. But during the Industrial Revolution, as new methods of producing bar iron with coke instead of charcoal were developed, charcoal too ceased to be a strategic commodity. A similar thing happened to salt.

The story of salt

Until the advent of canning, refrigeration, and other food preservation techniques, salt topped the list of strategic commodities. "Salt is so common, so easy to obtain, and so inexpensive that we have forgotten that from the beginning of civilization until about 100 years ago, salt was one of the most sought-after commodities in human history," wrote Mark Kurlansky in his book *Salt: A World History*.[1] The white mineral, which Homer called the divine substance, was used in multiple applications among them curing hides, separating silver from ore and, in the 18th century, manufacturing munitions. But what made salt a real strategic commodity was its unique ability to preserve food. Salt removes water from food and creates an inhospitable environment for harmful bacteria. Since the time of ancient Egypt, it was used to cure meat, fish and vegetables. This allowed people to store food for times of drought, famine, disease and other rainy days. It also enabled militaries to march across continents – as Napoleon Bonaparte observed "An army marches on its stomach" – to withstand long sieges and to send their navies on long expeditions. Salt was needed to maintain the health of soldiers and their horses. Roman soldiers were paid in salt; the word salary is derived from the Latin for salt. During Napoleon's retreat from Russia, thousands of French troops died from minor wounds because of lack of salt for disinfectants.[2] As a strategic commodity, salt had profound implications for the fate of empires. During the four hundred years of the Chinese Tang dynasty, half of the revenue of the state was derived from salt.[3] The Mayan civilization declined when its salt trade system broke down. And cities like Rome, Liverpool and Salzburg – Salzburg means "Salt Castle" and is located on the Salzach (from the German word Salz) river which until the 19th century was an important shipping route for salt – were founded in proximity to saltworks or salt trade routes. Wars were fought over salt, and colonies were built around it. Salt was a major ingredient in the production of herring, a fatty, high protein fish which for centuries was a staple of the European diet. In 1360, the Danes went to war with the Hanseatics over control over herring fisheries.[4] In 1652, the British Navy destroyed the Dutch herring fleet.

Wars were also fought *directly* over salt. In the 17th and 18th centuries, the British, the Dutch, the Swedish and the Danes were all engaged in a frantic search for islands with salt. They found them in the Caribbean where salt centers like Tortuga, Boa Vista and Turk Island enjoyed the same strategic importance as today's Persian Gulf oil emirates. Salt also shaped foreign policy and alliances. "To the British admiralty, the solution to a lack of sea salt was to acquire through war or diplomacy places that could produce it. Portugal had both sea salt and an important fishing fleet, but needed protection, especially from the French who were regularly seizing their fishing boats. And so England and Portugal formed an alliance trading naval protection for sea salt," wrote Kurlansky.[5] Replace the words "British", "Portugal" and "salt" with "American", "Saudi Arabia" and "oil" and you find yourself jumping from 18th century reality to that of our own days. Just like with petroleum, the desire of regimes to monopolize salt and tax their consumers was a great source of tension in the international system during all those centuries in which salt was a strategic commodity. Salt taxation was one of the catalysts that sparked both the American War of Independence and the French Revolution. In both cases, heavy handed policy by the crown triggered anger and social discontent. During the American Revolution, salt played such an important role that in June 1777, a special congressional committee was appointed to "devise ways and means of supplying the United States with Salt," and saltwork employees were exempted from military service.[6] The British Empire failed to learn the lessons of salt's importance for the common man and repeated the same flawed tax policy on salt in India. In 1930, Mahatma Gandhi and thousands of his followers performed their 240-mile march to Dandi, Gujarat, in order to defy British law by scraping salt. The conclusion of the march on April 6, 1930 sparked large scale acts of civil disobedience against the British salt laws by millions of Indians. "Today, thousands of years of fighting over, hoarding, taxing, and searching for salt appear picturesque and slightly foolish," Kurlansky wrote, "The seventeenth-century British leaders who spoke with urgency about the dangerous national dependence on French salt seem somehow more comic than contemporary leaders concerned with a dependence on foreign oil. In every age people are certain that only the things they have deemed valuable have true value."[7]

The reason salt is no longer a strategic commodity is *not* because we consume less of it or import less of it. On the contrary, both consumption and import dependence have been on an upward trajectory in recent decades. What has changed is the way we use salt. When salt was a strategic commodity, its main usage was in the food industry. Today, only three percent of our salt is used as food. Most of our salt, nearly 40 percent, is used in

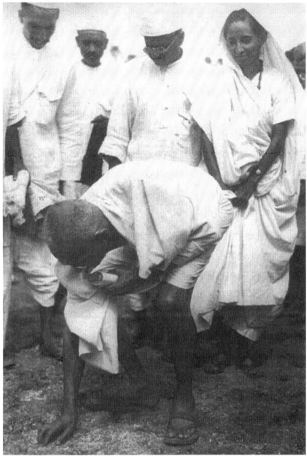

Gandhi scraping salt in Dandi, April 6, 1930

the chemical industry, mainly for chlorine and caustic soda manufacturing, and another 40 percent is used for highway deicing. So while we continue to use salt, it is no longer essential to the functioning of our society. Of course, if our salt supply were to be interrupted, there would be multiple direct and indirect implications for our lives, but society would not grind to a screeching halt.

A strategic commodity is therefore a commodity that underlies our very way of life and the functioning of our economy. It is an irreplaceable civilizational enabler which dominates our economic, geopolitical and cultural behavior. When something happens to knock a strategic commodity off its pedestal and importance shifts to other commodities, the structure of a society changes and so does its economic activity, the geopolitical balance of power among nations and their trade patterns.

One misconception about strategic commodities is that once their availability declines, increasingly hostile competition among consumers emerges, coupled with gradual empowerment of those who own their remaining reserves. Sometimes this is the case. The shift from wood to coal, for example, was a result of depletion of the forest areas surrounding rapidly developing urban areas. But often strategic commodities lose their status due to technological advances that simply deem them less essential. As Saudi Arabia's legendary Petroleum Minister Sheikh Zaki Yamani once pointed out: "the Stone Age did not end for lack of stones." The era of salt did not end because we ran out of salt. In fact, salt is almost a renewable resource. Naturally evaporated salt from seawater is found throughout the globe in sufficient locations and quantities to supply world demand forever. Salt depletion was not why salt lost its strategic status. It was canning and refrigeration, which offered superior ways to preserve food without the need to cure it, that did it. Salt is still found on almost every table and at our doorstep on icy winter days, and we are certainly not running out of salt, but it is no longer a commodity that underlies world events. Do we even know which nations our salt comes from? Do we have a salt dependence problem? Do we care? If we use too much salt, we are much more likely to hear about it from our cardiologist than from our president.

How coal lost its status?

Coal, too, did not lose its strategic status because of depletion. Coal has been in use since the 12th century, but it wasn't until 1600 that it essentially replaced wood as the dominant fuel of England, the biggest industrial power of the time. As Edwin Black described, coal became not only a matter of national security with squadrons of warships devoted to protecting the indispensable coal convoys, it was also an enabler to rivaling monopolies which held a tight squeeze on England's economy and accumulated inordinate power only diminished by England's brutal Civil War in the 1640s.[8] The massive rise in coal use occurred nearly two centuries later during the Industrial Revolution. Coal was a critical raw material for the chemical industry, for steel production, for medicines and, of course, it became the primary fuel for steam engines and power plants. Coal consumption doubled every decade between 1850 and 1890, and by 1900 coal provided 70 percent of U.S. energy.[9] When in the late 19th century electricity began to be generated in central power stations, coal became the world's most important strategic commodity. As John Maynard Keynes observed, "the German Empire was built more truly on coal and iron than on blood and iron."[10] In many European countries, the coal miner became a powerful symbol of social revolt and industrial strength as illustrated by Emile Zola's novel *Germinal*, based on the strikes in France's northern coal fields in the 1880s. Entire communities were formed around coal mining and shipping. Canals and railways crisscrossed nations in order to transport coal, and steam propulsion gave rise to powerful and fast ships that enabled the big empires to colonize the world. Today, it is still premature to talk about the end of coal, though Al Gore, Google founders and other enviro-visionaries strive to make the day near. Coal currently provides 40 percent of the world's electricity. It is cheap and abundant, and despite the environmental problems associated with its use it is likely to continue to play a major role in the power sector in the decades to come. The United States alone has nearly 300 billion tons of recoverable coal, roughly a quarter of the world's total, enough to last more than 250 years at current rates of consumption. But make no mistake. Coal is no longer a strategic commodity. While its use in absolute tonnage is on the rise, its status as a strategic commodity has been in decline for nearly a century. First, coal lost its hegemony in the transportation sector starting with the British Navy's historic decision just prior to World War I to shift its battleships from running on coal to oil. This was followed by the transition to diesel fueled locomotives. Then, its hegemony in the power sector was challenged. Coal became substitutable by other sources of electricity such as nuclear power, natural gas, hydro, geothermal, solar, wind and biomass. More and more countries are diversifying their electric generation portfolio away from coal, and in some countries the use of coal has become almost non-existent. Take France as an example. In April 2004, a quiet ceremony in the La Houve mine near Creutzwald in Lorraine, France, marked the end of the French coal mining industry – an industry that produced 60 million tons of coal and employed 300,000 people as recently as 40 years ago. This was the outcome of a decision taken ten years prior by the French government to phase out coal production. Instead, France has gone nuclear. France began beefing up its civil nuclear program after the 1973 Arab Oil Embargo. Over the next decade, the country built nearly two reactors a year. While the nuclear accidents at Three Mile Island and Chernobyl turned public opinion in the United States and many European countries against nuclear energy, in France nuclear power moved from strength to strength. France today has

58 nuclear plants which together meet 80 percent of its total electricity needs – and allow it to export power to Britain, Germany, Switzerland and Italy. For the French, uranium may be a strategic commodity, but coal certainly isn't.

The power of strategic commodities

When the first automobiles appeared on the road at the end of the 19th century, gasoline was only one of several fuel options. For decades, gasoline competed with alcohol fuels, electric cars and steam engine cars. In the early 20th century, Thomas Edison and Henry Ford were well on their way to going into business together to offer low cost electric cars. The Model T, which was produced by the Ford Motor Company from 1908 through 1927, was capable of running on gasoline or alcohol. A variety of reasons – from gasoline's high energy density and its decreasing cost, to corporate interests and politics (prohibition sealed the fate of alcohol as a viable motor fuel) – gasoline became the fuel of choice of the nascent American auto industry. As the United States set the standard in the automotive market, the rest of the world followed, and petroleum became virtually the sole source of transportation energy.

Because most of the world's oil reserves are held by countries that are highly unstable, corrupt, dictatorial and in some cases hostile to the United States and its allies, oil's monopoly in the transportation sector is one of the most destabilizing factors in the international arena today. For most of the 20th century, world oil supply was relatively uninterrupted, and the occasional conflicts and disruptions (the 1973 Arab Oil Embargo, the 1979 Iranian Revolution, and the 1990-91 Gulf War) were short lived and relatively contained. They certainly did not merit a fundamental change in our energy system. But in recent years, it has become increasingly apparent that the world is facing a "perfect storm" of security and economic problems, all directly linked to our oil dependence. On the eve of September 11, 2001, oil averaged less than $20 per barrel. A year later, it stood at $30. In July 2008, oil prices surpassed $145 and then collapsed to under $40 as the economic crisis hit. During the spike in oil prices, oil-rich countries such as Saudi Arabia, Iran, Russia and Venezuela were awash with petrodollars and able to actively oppose America's foreign policy goals – thwarting radical Islam, spreading freedom and democracy around the globe and preventing the proliferation of nuclear weapons – and pose a challenge to global security.

Want to stop nuclear proliferation? Forget about it. While the United States and the European Union are trying to forge a diplomatic strategy to halt Iran's nuclear program, Iran advances its anti-Western agenda and its pursuit of nuclear weapons technology secure in the knowledge that with 10 percent of the world's known oil reserves and the world's second-largest natural gas reserves, it is virtually immune to sanctions. If Iran does succeed in becoming a nuclear power, the long term consequences could be far more severe. A nuclear Iran will not only be a threat to U.S. interests in the Middle East – Iran has been meddling in Iraq and its President Mahmoud Ahmadinejad is a passionate advocate of the destruction of Israel – but it also guarantees that other Middle Eastern countries follow suit. Many regional actors including the Gulf Cooperation Council (GCC), Yemen, Egypt, Jordan, and Morocco have already stated their intention to develop nuclear capabilities. Of course all of these countries explain that the development of nuclear power is exclusively "for peaceful purposes." And who would suspect that those peace loving monarchs mean otherwise? But considering the history of miscalculation and erratic behavior by some Middle East regimes and the risk of a nuclear country taken over by a Taliban style regime, a nuclear Middle East is a spectacularly bad idea.

Whether or not the strategic status of oil will eventually enable the Iranian regime's genocidal ambitions regarding Israel is not known. But it has already allowed one genocide to take place. The world's worst current humanitarian crisis is currently taking place in Darfur, where 200,000 people have been killed and 2.5 million have been displaced over the past five years. While the United States and other governments urge the United Nation's Security Council to impose harsh sanctions against Sudan, which is believed to own Africa's largest unexploited oil resources, China, Sudan's largest trading partner, is on the other side of the issue. China's thirst for Sudanese oil turned Beijing into Khartoum's largest supplier of small arms and its diplomatic protector. For China, oil is clearly thicker than blood.

The flow of petrodollars to major oil exporters is a counterweight to another U.S. strategic objective – democracy promotion. Studies show that countries rich in easily extracted and highly lucrative natural resources that do not have well-developed democratic traditions, do not sufficiently invest in education, productivity, or economic diversification. In addition, such resource-rich governments do not feel obligated to be accountable or transparent to their people, and they deny them representation. They also have no imperative to educate women or to recognize their inherent entitlement to equal rights. With few exceptions, oil exporting countries demonstrate an abysmal record in terms of human rights, political stability, and compliance with international law. Only 10 percent of the world's oil is concentrated in countries ranked "free" by Freedom House. In authoritarian countries highly dependent on oil and gas for their income, such as Myanmar, Sudan, Azerbaijan, Uzbekistan, Kazakhstan, Angola, Nigeria, Chad and Russia, freedom has been in retreat since oil prices began their climb.

Bombs and barrels

Winning the war on radical Islam in a world dominated by oil is an oxymoron. Despite promises by Middle Eastern governments to stop terrorist financing, nearly a decade after September 11, wealth generated by the region's oil rich countries continues to flow to terrorist organizations and organizations promoting radical Islam. While the U.S. economy hemorrhages hundreds of billions of dollars every year to pay for imported crude, oil-producing nations such as Saudi Arabia and Iran that are sympathetic to, and directly supportive of, radical Islam are on the receiving end of staggering windfalls. In 2005, OPEC's net export revenues were $473 billion, more than double the amount generated in 2001. In 2008, the figure surpassed $1 trillion. An undetermined portion of that money finds its way – through official and unofficial government handouts, charities and well-connected businesses – to organizations promoting militant Islam. Stuart Levy, the U.S. Under Secretary of Treasury for Terrorism and Financial Intelligence, said: "If I could snap my fingers and cut off the funding from one country, it would be Saudi Arabia." Furthermore, oil's strategic value requires increased Western military presence in the Persian Gulf to keep the sea lanes of communication open and provide protection to key oil producers. This presence strengthens the xenophobic and anti-Western sentiment among jihadists and increases their motivation to fight the "infidels." At the same time, a continuous infusion of petrodollars to radical Islamic educational institutions creates a new generation of radicalized youth, making reconciliation between the West and the Muslim world more difficult to achieve. This cycle can only be broken through massive political reforms that the oil regimes currently seem to resist. Furthermore, determined to weaken the Western economy, the jihadists have made attacking oil, which they call "the provision line and the feeding to the artery of the life of the crusader's nation," a central part of their plan to bring about U.S. economic decline. Over the past five years, attacks on oil facilities, primarily in Iraq, have denied the global oil market 1-2 million barrels per day, not including thwarted investment in capacity expansion. Had this oil been in the market, the price per barrel would have dropped significantly and tens of billions of dollars that ended up going to the Middle East would have stayed at home, creating jobs and investment opportunities for hard working Americans.

Oil's monopoly in the transportation sector and the influx of money it provides oil exporters emboldens bad actors. Russian control over a large part of the world's oil and gas market allows it to play hard-to-get on Iran, jail opposition leaders, bully its European neighbors, and work to roll back democracy in former Soviet republics such as Ukraine and Georgia. Russia's repeated use of energy as a political weapon as well as its August 2008 attack on its neighbor Georgia may be harbingers of a new Cold War. High oil prices also undermine America's position in its backyard. Petrodollars lubricate the so-called Bolivarian revolution led by the anti-American president of Venezuela, Hugo Chavez, who is using his windfall oil wealth to buy political influence in the Western Hemisphere and consolidate an anti-U.S. bloc. U.S. diplomacy is further complicated by the thirst for energy of emerging countries like China and India, which are becoming increasingly dependent on the very same countries the United States is trying to rein in. The growing appetite of developing Asian powers not only plays into the hands of the aforementioned rogue producing nations, but also feeds what could become a global competition for control of energy resources. Rogue nations like Iran and Sudan can now buy themselves the support of a third of humanity – not to mention the protection of Chinese veto power on the U.N. Security Council – by signing energy deals with China and India.

Holding the lion's share of the world's reserves, oil producing governments constrain supply by practicing resource nationalism, sticking to production quotas and obstructing international companies from investing in their territories. The main casualties of this transfer of wealth are the world's poorest nations, some still carrying debts from the 1970's oil shocks, which are often forced to buy oil at inflated prices. This has profound implications for global security, driving regional unrest, increasing poverty, and nipping in the bud progress towards democracy. In the years to come, when the world economy heals and oil prices bounce back, the economic imbalance between producers and consumers will grow by leaps and bounds. OPEC's market capitalization based on its proven reserves is roughly equivalent to the value of the world's total financial assets – stocks, bonds, other equities, government and corporate debt and bank deposits – even at their current distressed level. When oil prices rise, this monumental wealth potential translates into massive buying power for sovereign wealth funds controlled by the regimes of oil countries, which enable these regimes to acquire control over critical nodes of the global economy and pose a challenge to U.S. economic sovereignty.

Indeed, oil's strategic status is a poison pill for everything the United States is trying to accomplish abroad. The person who perhaps knew this best is former Secretary of State Condoleezza Rice, who in testimony before the Senate Foreign Relations Committee, on April 5, 2006, admitted: "We do have to do something about the energy problem. I can tell you that nothing has really taken me aback more, as Secretary of State, than the way that the politics of energy is [...] 'warping' diplomacy around the world. It has given extraordinary power to some states that are using that power in not very good ways for the international system, states that would otherwise have very little power."[11] But this "something" Ms. Rice suggested we must do about our energy problem, becoming more energy independent, is perhaps the most abused, misused and confused term in our political discourse. If we are to move toward energy independence, we should first understand what it really means, and even more important, what it doesn't.

⌘   ⌘   ⌘

2
WHAT IS ENERGY INDEPENDENCE ANYWAY?

It seems that every politician has a speech ready about the need to become energy independent. Twenty four of the 34 State of the Union addresses since the oil embargo of 1973 have proposed solutions to take us there. In the midst of the 2008 presidential campaign, Senator John McCain declared, "We need energy independence," while his opponent, Senator Barack Obama, promised "serious leadership to get us started down the path of energy independence." Public opinion poll after poll show that Americans are fully on board, viewing energy independence as a top national priority. Yet, despite the popular appeal of energy independence, in many circles the concept is met with strong skepticism; in some cases outright contempt. One can understand why the Saudi Oil Minister refers to energy independence as a "fallacy." It is also understandable why a 2007 report by the National Petroleum Council, a federally chartered group that offers advice from the oil and gas industries to the government, calls energy independence "unrealistic"[1]; why a book by Robert Bryce, a former fellow at a think tank funded in part by energy interests, described energy independence as "a hogwash," a "gusher of lies" and a "dangerous delusion"[2] or even why the dean of energy industry historians and Pulizer Prize winner Daniel Yergin whose consulting firm Cambridge Energy Research Associates advises Big Oil, wrote in *Foreign Affairs*, that "real energy security requires setting aside the pipe dream of energy independence and embracing interdependence."[3]

Less understood is the broad collection of distinguished Americans who dismiss and rebuff the concept. John Deutch, former Director of Central Intelligence said "energy independence is not a constructive idea." Former Secretary of Defense and Energy, James R. Schlesinger, called it a "forlorn hope,"[4] and Michael Kinsley, a columnist for *Time Magazine* called it "a muddled concept and a red herring."[5] A report by the conservative Hudson Institute starts by saying that "no energy policy discussion is worth having if it does not start with a hard fact: there is no practical alternative to continued U.S. dependence on crude oil imported from nations that at best do not wish us well, and at worst wish us serious harm."[6] Jerry Taylor from the libertarian Cato Institute wrote that "characterizing the idea [of energy independence] as ridiculous is charitable."[7] Frank Verrastro and

Sarah Ladislaw from the centrist Center for Security and International Studies (CSIS) referred to energy independence as "a misguided quest,"[8] and Robert Manning from the Atlantic Council called it "a seriously flawed idea."[9] "Talk of energy independence is ridiculous and may not even be a worthwhile goal," contend the experts of the Baker Institute for Policy Studies at Rice University.[10] Flynt Leverett, senior fellow and director of the Geopolitics of Energy Initiative at the New America Foundation, a progressive Washington think tank dedicated to "bring *exceptionally promising* new voices and new ideas to the fore of our nation's public discourse," remarked: "Whether in the name of national security, whether in the name of climate change, whether in the name of anything else, energy independence is a myth. [...] If we waste time on illusions of energy independence [...] in 20 years we will be worse off [...] than we are today."[11] General Chuck Wald, former deputy commander of US forces in Europe, proclaimed that energy independence would be "the worst thing to happen to America."[12] And a Council on Foreign Relations task force went so far as to accuse those promoting energy independence of "doing the nation a disservice."[13] All those intellectual pearls coming from the best of the nation's think tanks, talking heads and Washington insiders from all parts of the political spectrum have infected the bloodstream of the U.S. government. In July 2008, a consultant to the Department of Energy shared with us an internal email sent to the Department's employees from Adam Ingols, then chief of staff to the Secretary of Energy Samuel Bodman: "For future reference we should never use the phrase 'energy independence' – the notion of which is not only an impossibility, but potentially detrimental to U.S. and global interests/markets. Instead, we seek to 'reduce our dependence on foreign sources of energy." We wonder which "interests" he was referring to.

These wall-to-wall denunciations of energy independence come for good reasons. For some, the idea is a challenge to an existing order on which they thrive intellectually and/or financially. Indeed, an army of well funded defenders of the status quo has had years to entrench itself in the suites of the District of Columbia's K Street. Others on the left who are primarily concerned with combating climate change view energy independence, which for many in America's political culture is synonymous with increased production of domestic oil, as a distraction from broader conservationist goals. There are also those who fear that the talk about energy independence will cause producers in the Middle East and Russia to cut back on investment as future demand for their product becomes less certain.

And then there are those who share the vision of a happy global community in which nations should become more, rather than less, dependent on each other. To their ears, energy independence has an insufferably jingoistic and isolationist ring. Verrastro and Ladislaw for example called for "a much more *sophisticated* approach to energy policymaking, one that more fully appreciates the interdependencies of global markets, the complex nature of energy security, and the need to manage the trade-offs inherent in energy policy decision making."[14] Unsophisticated us accept the benefits of economic interdependence, but we are still not convinced that it guarantees peace and economic stability, certainly not energy security. Sometimes it does; sometimes it doesn't. World War I broke among the most economically interdependent countries. Despite high trade levels, in 1913-14 German leaders decided to attack to ensure long-term access to markets and raw materials. In the 1930s, the two most aggressive states, Germany and Imperial Japan, were also the most highly interdependent despite their efforts towards autarky, relying on other states for critical raw materials. In fact, Japan had a much higher level of economic interdependence with other countries than it did in the 1920s, but nonetheless it embarked on aggressive imperialism that ended in Pearl Harbor followed by two mushroom clouds. It is also worth noting that today, as Europe and Russia are practicing the doctrine of energy interdependency, Europeans are less secure than ever when it comes to their energy supply. But what is perhaps the main reason for the pundits' distaste of energy independence is their narrow and literal interpretation of the concept. Critics of energy independence correctly point out that the United States is not at all dependent on the Middle East. Only one eighth of our oil originates in the Persian Gulf. The two largest suppliers of crude to the U.S. market are Canada and Mexico – neither exactly known as a foe of the United States. This true but highly irrelevant argument reflects the biggest misunderstanding among the naysayers of the meaning of energy independence: they all define "independence" as essentially "autarky" – i.e. complete self-sufficiency, or not importing oil even though we remain dependent on it. In their *Washington Post* op-ed, Roger Sant and Michael Kinsley articulated this common, yet flawed, view of what energy independence is: "when we produce as much energy as we consume: net energy imports (basically, oil imports) of zero."[15]

Such a simplistic definition of energy independence as self reliance or "energy isolationism" captures none of our thinking. The point of independence is not to be an economic hermit or substitute cheap Middle

Eastern imports with oil from domestic sources or from our nearby neighbors and somehow manage to insulate ourselves from the world oil market. The world marketplace doesn't work like that. Oil is a *global* commodity and the price of oil is set globally, not locally. Several years ago, when the United Kingdom was still a net oil exporter, British truckers went on strike because fuel prices soared. It didn't help them a bit that their country was then self sufficient in oil. Think about the oil market as a huge bathtub into which producers pour oil and from which consumers sip oil with different intensities and straw sizes. It doesn't really matter if the United States stops buying foreign oil from countries with which it is at odds. We don't buy oil from Iran, but that doesn't stop the Islamic Republic from selling its oil to other clients, from calling on OPEC to cut production and threatening to stop the supply of oil through the Straits of Hormuz. Any disruption in Iranian supply affects prices for everybody, not just for countries that import oil from Iran.

What makes oil unique?

What the critics ignore is oil's uniqueness in the family of commodities and goods. They fail to recognize what makes oil a strategic commodity second to none. This makes them treat the oil market just like any other market, a mistake that afflicts even some of the most enthusiastic advocates of reducing oil dependence. One of them, for example, is Andy Grove, the legendary former chairman and chief executive officer of Intel Corporation, the world's largest producer of microchips. Just like many other Silicon Valley dwellers, in recent years Grove has become an enthusiastic advocate of electric cars, and he even devised an interesting plan to deploy a large number of electric cars by providing incentives to convert old ones. Yet, when it comes to energy independence, he is less sanguine, calling the concept "a faulty goal." In an article in *The American Magazine*, he explained what he thought is wrong with energy independence: "The United States became more and more integrated into a global economy, where goods, information, and oil move unimpeded across national boundaries. Countries around the world produce energy if they can, and buy on the world market what they need beyond their own production. Oil flows toward the highest bidder, just like all other goods. Consequently, talking about 'independence' in terms of one product in an otherwise seamless global economy is a contradiction."[16]

Just like all other goods? Let's compare oil to microchips, something Grove certainly knows a lot about. No doubt both are essential to the world economy. Oil enables the flow of people and goods while semiconductors enable the flow of information. But this is as far as the similarities go. Oil is not like any other good or commodity, first because of its aforementioned strategic status which allows it to dominate the world's economy, second, because of the nature of the countries that hold its reserves and, third, because it is not traded freely but rather controlled by a cartel. Any other commodity fits this description? The top ten producers of semiconductors are publicly owned companies. The top ten producers of oil are governments, and not the best of them. Unlike Saudi Arabia, Iran, Venezuela and the other major oil exporters who convene in the OPEC headquarters in Vienna every few months to manipulate oil prices, no such cartel exists in the semiconductor industry. On the contrary, Toshiba, Texas Instruments and Intel compete *against* each other rather than collude *with* each other.

Imagine if the only thing we could eat for breakfast was cornflakes and that the main corn producing countries colluded to form a cartel, holding us over a bushel. Luckily that's not the case. But even if it were, the impact on our lives would be minimal. The reason: We have options. Rice Krispies or granola can make our morning just as well. Therefore, when a foreign country dominates a non-strategic commodity there should be no reason for alarm. Unless you are addicted to Nutella, the fact that 70 percent of the world's hazelnuts are grown in Turkey should not be a source of concern even if for some reason a radical Islamic regime were to take over Istanbul and decide to stop supplying the world with hazelnuts. Few people know that Sudan, a country that for years has been committing some of the world's most heinous crimes against humanity, controls the world supply of a commodity that is used by all of us. Sudan supplies the world with about 80 percent of its gum arabic, an emulsifier made from the acacia tree that adds to the fizz of sodas. In a June 2007 press conference, Sudanese Ambassador to the U.S. John Ukec Lueth raised a bottle of Coca Cola and warned that his country would stop supplying the United States with gum Arabic in response to U.S. sanctions against the government of Sudan for its involvement in the Darfur genocide.[17] Despite the fact the United States has among the world's highest per capita soda consumption, the ambassador's threat provoked chuckles rather than fear. But if Sudan were in control of 80 percent of a strategic commodity like oil, the response would have been very different. The thing to remember is that oil is not corn, gum arabic or a silicon chip and therefore it writes its own rules. To be more precise, it is the oil exporters that write the rules. If we cannot participate in writing the rules we might as well change the game altogether.

What *is* energy independence?

Changing the game is what energy independence is all about. Energy independence is our way of freeing ourselves from the control of a club of unfriendly countries that happen to sit on the lion's share of the world's petroleum reserves and therefore control the global economy. It also means freedom to act without fear of economic repercussions. For decades, the need for oil forced the United States to enter brutal friendships with some of the world's worst despots, supplying them advice and state-of-the-art weapons which enabled them to stay in power while oppressing their people, a policy that has boomeranged against the United States.

One example was America's forgiving, oil-dictated treatment of the Shah of Iran despite his corruption and human rights abuses. When the Shah fell, the Iranian people responded with an outpouring of anti-Americanism felt to this date. America's support for the House of Saud has produced a similar sentiment. In the 64 years since the historic meeting aboard the USS *Quincy* in Egypt's Great Bitter Lake between an ailing President Franklin D. Roosevelt and King Abdul Aziz ibn Saud, the founder of the Saudi monarchy, the United States has treated Saudi Arabia with considerable deference. This deference began when FDR's decided to refrain from smoking in the King's presence (when he could not hold out anymore, he excused himself and went to the ship's elevator to smoke a cigarette), continued with President Bush's hand-holding and President Obama's genuflection before Abdul Aziz's son, King Abdullah during the G20 summit in April 2009. These symbolic acts of self-denigration mask a political obsequiousness that seriously undermines our national security. Wrongly labeled as the "War on Terrorism," the current war in which we are embroiled is in essence against an ideology, and this ideology is radical Islam (and the Obama Administration's decision to banish the phrase by no means rectifies this misrepresentation of the nature of the war.) There is Sunni radical Islam propagated by Saudi Arabia and there is Shiite radical Islam promoted by Iran. They may hate each other, but they both hate us more. Under the leadership of Mahmoud Ahmadinejad, a fanatic Holocaust denier who believes that the return of the 12th Imam can only happen through the agonizing scream of millions of burning people, Iran has turned into a medieval society where gays are executed by slow hanging, women are harassed by police for not covering themselves, opposition members endure inquisition-style torture and petty criminals are mutilated and amputated. In the country that holds so much of the world's oil and gas, death by stoning is an acceptable practice, and if you'd like to participate in this death-fest, heed the government's instructions about the size of stones to be used: not too big so the prisoner doesn't die quickly, not too small so he or she doesn't survive.

No less primitive behavior can be seen in Saudi Arabia. Despite the fact that 15 out of the 19 terrorists on September 11 were Saudi nationals, the Saudis continue their sponsorship of militant Islam to the tune of billions of dollars annually, which enables the Wahhabi sect to establish religious schools and institutions throughout the world. Thanks to the Kingdom's policies, almost 100,000 young boys graduate every year from some 12,000 madrassas, Koranic schools where students are required to recite the Koran by heart and are brainwashed to hate Americans, Brits, Christians, Jews, atheists and other "infidels." Each of those 12,000 factories of mini-weapons of mass destruction can produce monsters of the kind that carried out the 9/11 attacks or the terror attacks in Mumbai, Madrid, London, Bali or Tel Aviv. Many of those killing machines would happily detonate a nuclear weapon in the middle of a large American or European city if he or she were only given the chance to do so. To make matters worse, Saudi Arabia's next in line for succession, Crown Prince Nayef, is a fanatic Wahhabi who publicly rejected the notion of Saudi involvement in September 11, suggesting that Israel's intelligence service, Mossad, was behind the attacks. This is the same Prince Nayef who headed the religious police that a few years ago shoved little girls back into the burning school they were fleeing, because they didn't have veils on their heads. The thought of this fanatic mind controlling the world's oil supply is frightening. Iran and Saudi Arabia, two countries that if not for their oil and gas would be treated as Taliban countries, are treated by the world with forgiveness. True, since it revealed its intentions to develop nuclear capabilities and wipe Israel off the map, Iran has been somewhat isolated by the West. But this didn't stop Western companies from continuing to do business with the Islamic Republic, and when in April 2009, the United Nations hosted its so-called conference against racism in Geneva (aka Durban II), the guest of honor was the up and coming Hitler of the 21ˢᵗ century – the Holocaust denier and advocate of genocide Mahmoud Ahmadinejad. In fact, Ahmadinejad was the *only* head of state to attend Durban II.

This policy of forgiveness, indifference and, at times, subservience, toward our ideological adversaries can be explained only by our desperate need for cheap oil and gas. In February 2005, President George W. Bush conceded that "The policy in the past used to be, let's just accept tyranny, for the sake of [...] cheap oil [...] and just hope everything would be okay? Well, that changed on September the 11th for our nation. Everything wasn't okay. Beneath what appeared to be a placid surface lurked an ideology based upon hatred." This lesson is being internalized in each place terror strikes. After the July 2005 London subway bombings, that city's mayor at the time, Ken Livingstone, denounced nearly a century of foreign policy on the Middle East as coddling tyrants and ignoring violations of human rights "because of the Western need for oil." But despite all that, our governments' response to the spread of Wahhabism has ranged from expressions of concern to softly-softly asking the Saudis to tone down the vitriolic rhetoric expressed by their state-sponsored clerics. Quick as we were to bomb Saddam Hussein's laboratories of weapons of mass destruction, we have done virtually nothing to address the threat of Saudi-sponsored human guided bombs.

Eagle's head in Saudi sand

Our deference to the Saudis apparently knows no boundaries. Most Americans heard about the travesty of allowing a plane full of Saudis, including members of the bin Laden family, to sneak out of the United States immediately after September 11 in order to avoid FBI questioning. We have also learned about extensive financial support for al Qaeda and other extremist groups by members of the Saudi royal family, and about the missing 28 pages in the congressional report that dealt with Saudi Arabia's role in the September 11 attacks.[18] But fewer noticed how U.S. officials whitewashed the scandalous Saudi policy of allowing Saudi jihadists to cross the border to Iraq in order to fight and kill U.S. soldiers during the first few months of the Iraq War. Hearing administration officials comment on terror attacks against coalition forces carried out by "foreign fighters infiltrating from neighboring countries," one might have thought that Iraq borders only two countries: Syria and Iran. In September 2003, U.S. administrator in Iraq Paul Bremer told reporters that foreign fighters captured by U.S. forces were mostly from Syria and Iran. A month later, in response to multiple bombings in Baghdad, President Bush alluded to the fact that Syria and Iran were the source of foreign fighters: "We expect them to enforce borders, prevent people from coming across borders." Defense Secretary Donald Rumsfeld, too, said that between 200 and 300 fighters believed to be foreigners had been captured, with a "high percentage" from Syria. These lines were echoed a number of times by State Department Spokesman Richard Boucher who promised that "the issue of people coming across the border from Syria and Iran remains a concern of ours." Without belittling the destructive role of Syrian and Iranian fighters, one group of nationals from a two-word country always stood out in its absence from official condemnations. Yet, all along there was ample evidence to suggest that the largest contingency of foreign fighters came from Saudi Arabia, crossing the 475-mile-long border with Iraq, and not only that, but among them were some of the most dangerous terrorists U.S. forces faced. According to a 2003 *Financial Times* report, up to 3,000 Saudi al Qaeda sympathizers went missing in Saudi Arabia and are likely to have heeded calls for holy war by Osama Bin Laden to cross the Saudi-Iraqi desert border. A 2005 analysis by NBC's investigative unit revealed that 55 percent of the foreign fighters were Saudis.[19] In September 2007, the U.S. military discovered documents and computer data that belonged to al Qaeda after conducting a raid in Sinjar in western Iraq. The more than 750 personnel records obtained at the raid site showed that 41 percent of the foreign fighters in Iraq originated from Saudi Arabia.[20] Despite evidence that Saudi Arabia was the source of the most dangerous and suicidal Arab terrorists, Washington still refuses to publicly discuss the role of Saudis in the Iraqi insurgency or refer to Saudi Arabia in the same tone or wording used for other uncooperative, rogue neighbors. Furthermore, throughout the first years of the war there was no demand that the Saudis beef up their military presence along the Iraqi border. Nor has there been any special effort by the Saudi government, which claims to be a close ally of the U.S, to block the flow of jihadists to Iraq. Instead, the Saudis, through their then spokesman in Washington, Adel al-Jubeir, were quick to pass the blame to the United States, announcing that if extremists are getting across the border, it is the responsibility of U.S. forces to stop them. With an active military of 150,000 men and with no enemy threatening their country since the removal of Saddam Hussein, the Saudis were well equipped to seal the border as U.S. troops faced the monumental task of controlling tribal lands roughly the size of California. But they didn't.

The Saudi regime's complicity in the actions of its citizens in Iraq is similar to its complicity in the actions of al Qaeda prior to and following September 11. In both cases, the United States paid the price. The Saudis' welcome message to President Barack Obama upon taking office was chilling. With the backdrop of Israel's operation in Gaza against the terrorist group Hamas, Prince Turki al-Faisal, former Saudi ambassador to the UK and the U.S. and the man who during his tenure as director of the Saudi intelligence essentially ran Osama bin Laden's network in Afghanistan,[21] told a gathering of U.S. and Gulf experts in Riyadh in January 2009: "The Bush administration has left you (with) a disgusting legacy [...] Enough is enough, today we are all Palestinians and we seek martyrdom for God and for Palestine."[22] In an equally outrageous op-ed in the *Financial Times* a few days later, Turki, a Georgetown University classmate of President Bill Clinton, sent a message to Obama and to his former classmate's wife, now America's top diplomat: "If the U.S. wants to continue playing a leadership role in the Middle East and keep its strategic alliances intact – especially its 'special relationship' with Saudi Arabia – it will have to drastically revise its policies vis-à-vis Israel and Palestine."[23] He also threatened that without a change in U.S. foreign policy, Saudi Arabia might join Iran in a joint Sunni-Shiite jihad. This is not the first time Saudi Arabia has used its position as the world's leading oil supplier as a tool of coercion and intimidation. It is also not the first time the U.S. has maintained diplomatic silence in response to a provocation and challenge by its so-called ally. As Thomas Friedman once wrote "addicts never tell the truth to their pushers."[24]

When addicts speak out

What happens when a drug addict *does* tell the truth to his pusher? Or what happens when he comments on his behavior or – horror – criticizes him for beating up his wife or his employees? Switzerland can teach us something about this. On July 15, 2008, Swiss authorities arrested Hannibal Qaddafi, the son of Libya's leader Muammar Qaddafi, at a Geneva hotel for beating his two domestic employees. Hannibal has a rich criminal record in several European countries which includes reckless driving, illegal possession of a firearm and alcohol abuse. In 2005, he was sentenced in France for causing bodily harm to his pregnant girlfriend. Beating one's employees may be acceptable in Libya but for the law abiding Swiss, Hannibal's actions were intolerable. What was Daddy Muammar's response? He cut oil supply to Switzerland (Libya supplies some 20 percent of Switzerland's oil) demanding that Swiss authorities drop the charges and apologize.[25] The Swiss dropped the charges and allowed Hannibal to leave the country after a settlement, but refused to apologize. Qaddafi renewed his pledge to halt oil shipments, and asked all other OPEC nations to join in an international embargo. At the same time, Libya threatened to pull $7 billion in deposits from Swiss banks. A small example of what happens when an addict dares speaking out.

Let us be clear, energy independence will not stop terrorism and will probably not prevent rogue regimes from obsessively pursuing nuclear weapons. It will also not stop the swagger and bullying of oil rich dictators and kleptocrats. The cases of North Korea, Iran and Iraq under Saddam Hussein clearly show that ultra-ruthless regimes are willing to defy the world and starve and sacrifice millions of their own people in order to get the bomb. Energy independence would certainly not make these totalitarian actors reform their behavior. What energy independence *might* do is shrink the pool of resources available for them to support the industry of hate they have created. More importantly, it would enable the United States and its allies to draw a line in the sand and become more assertive in their relations with oil producing countries. Energy independence would spare us the humiliation of seeing the spectacle of our president going to Saudi Arabia – a medieval kingdom where women cannot drive a car, travel abroad by themselves, participate in the Olympic games or even work in a lingerie shop – to ask for oil as President George W. Bush did twice in 2008. Energy independence means not having to see our financial system increasingly controlled by sovereign wealth funds owned by some of the least transparent, most corrupt and human rights abusing governments in the world. Energy independence would permit our writers and publishers to report on Saudi human rights violations and support for terrorism without fear of expensive libel suits. Energy independence means restoring our freedom of action and our ability to operate on the world stage without fearing economic repercussions. The whole point of energy independence is to counter the ability of unstable and hostile regimes to influence U.S. policy and economic life by controlling the international flow of oil, and, in the case of Europe, natural gas. If we want to end dependence on the whims of OPEC's despots, the substantial instabilities of the Middle East, and the indignity of paying for both sides in the war on radical Islam, we must define oil "independence" sensibly – as Webster's Dictionary says, "not being subject to control by others," or in our case, doing whatever is necessary to avoid oil being the instrument of despotic leverage and foreign chaos. Independence means being a free actor by reducing the role of oil in world politics – turning it from a strategic commodity into merely another thing to sell. In other words, we must become independent – not just of imported oil, but of oil itself. Does this mean that we cannot use oil or import any? Of course not. Oil is a useful commodity that just like salt, tin and coal has thousands of applications. It already has competition from natural gas in industry and from natural gas and electricity for home heating. But in transportation, it has no competition, and its monopoly on transportation gives intolerable power to OPEC and the nations that dominate oil ownership and production. This monopoly must be broken.

To tell us that in following this path we are doing a "disservice to the nation" and should resign ourselves to oil dependence is like telling us we should not urge an alcoholic to stop drinking, but should rather impress upon him the health advantages of red wine, said James Woolsey. Such defeatist resignation also ignores the fact that in the electricity sector, the mission has already been accomplished. Remember President Jimmy Carter during the oil crisis in 1979, giving his famous "fireside chat" on energy in his yellow cardigan, urging Americans to save electricity by lowering their thermostats to 65 degrees during the day? It took us just one decade to wean the electricity sector from oil. Today, unlike the Europeans whose electricity generation is dependent on imported natural gas from Russia, the U.S. is almost self reliant for its power generation. It owns a quarter of the world's proven coal reserves; it operates more than 100 nuclear reactors; it has untapped natural gas reserves, and its system of rivers and dams produces hydroelectric power that meets nearly 7 percent of its electricity needs. Hardly any oil is used for power generation. The United States has no worry of Russian cuts of gas supplies and it is free to pursue its foreign policy in Eastern Europe without fearing that the light in Seattle or Washington DC will go out. Can we achieve something similar in the transportation sector, where American cars and trucks still gulp oil-based fuel? We sure can – and we should.

⌘   ⌘   ⌘

3
PLANS APLENTY, MOSTLY EMPTY

Despite the broad public understanding of the dangers of growing oil dependence and the urgent need for policy response at the international, federal and state levels, America's energy policy still suffers from institutional paralysis. The public discussion on the issue is marred with myths, fallacies, exaggerations as well as political and technological hype. This is caused partly due to partisan bickering but mostly, as we explained before, due to a poor definition of the energy problem. A century of domination by petroleum has led Americans to accept oil's monopoly over the transportation sector as a fait accompli. As a result, instead of addressing this monopoly as a problem to be solved, the focus from a domestic policy perspective has been on policies that increase either the availability of petroleum or the efficiency of its use. This has led to a public discourse that is overly focused on solutions that are politically contentious like domestic drilling and increasing mandatory fuel efficiency standards and by and large tactical rather than strategic or in the case of solar, wind and nuclear power, irrelevant to the problem, as very little electricity is generated from oil.

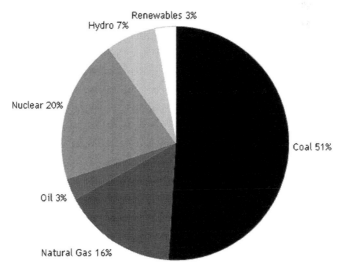

Sources of Electric Generation in the U.S., 2008

Four common policies have dominated the discussion of oil dependence. First, there has been great focus on an increase in domestic production in places like the Arctic National Wildlife Refuge (ANWR) and the Outer Continental Shelf (OCS). Second, as the Report of the National Energy Policy Development Group (the Cheney Report) and a Council on Foreign Relations Independent Task Force Report advocate, there is a call to increase the supply of oil from sources outside the Persian Gulf.[1] Third, a great deal of attention has been given to increased fuel efficiency of passenger vehicles as a way to reduce oil use. And finally, there has been a growing movement to develop alternative energy sources like natural gas, nuclear power, solar power or wind power as scalable substitutes to oil. The reality is that because those policies do nothing to challenge the status quo of cars as petroleum-only platforms, their implementation, even the implementation of all four of them, would have a negligible impact on oil's status as a strategic commodity and its domination of the transportation sector. Let's examine why.

Drill here, drill now, pay later

Of the four policy proposals, domestic drilling has caused the most acrimony and political gridlock. Liberals and environmentalists who, by and large, oppose tapping into America's oil reserves in Alaska and offshore invoke the need to protect pristine lands and coasts, urging more focus on policies to constrain demand. Conservatives and business interests see supply side solutions as a much more appealing way to deal with the problem. At the height of the 2008 oil crisis, when in some parts of the country gasoline was edging close to $5 a gallon, drilling became almost a new national religion. Proponents of this policy assume that if we only drill deeper, further and faster we can become energy independent. One of the most vocal drilling advocates is former Speaker of the House Newt Gingrich, one of the towering figures in the conservative movement. Only a few months after appearing with Speaker of the House Nancy Pelosi in a television ad campaign on behalf of Al Gore's Alliance for Climate Protection, warning Americans about global warming, Gingrich was out to redeem himself in the eyes of many angry and betrayed conservatives. He assumed the role of poster boy for the drilling movement. Through his nonprofit group American Solutions, the Driller-in-Chief mounted a campaign called "Drill Here, Drill Now, Pay Less!" which collected

1.5 million signatures on a petition calling for drilling. He even declared a video contest on why America "must adopt a 'Drill Here, Drill Now, Pay Less" approach." The prize: free gas for a year. Chuck Norris created a YouTube video supporting the movement, and country music singer Aaron Tippin dedicated a song whose chorus goes like that: "Drill here, drill now How 'bout some oil from our own soil that belongs to us anyhow No more debatin' we're tired of waitin' everybody shout out loud Drill here, drill now." With Alaska Governor Sarah Palin's entrance onto the national political scene as John McCain's vise presidential candidate, the drilling movement gained even greater momentum. The chant "Drill, Baby, Drill!" was lauded by millions in rallies across the nation. Drilling has become the Republican call to arms when it comes to energy policy. It became a code word that encapsulated their faith in the free market, the power of technology and American industriousness, their distaste of anything liberal or environmental, and, few would admit, their subconscious fascination with masculine symbols. (Real men drill, girly men conserve.) Throwing bones to the base during the pre-election season, the focus on drilling was a smart move, though in the final count it did not help Republicans win the elections. But as a linchpin of U.S. energy policy the "Drill Baby Drill" slogan, as Thomas Friedman pointed out, just makes America look stupid. "An America that is focused first and foremost on drilling for oil is an America more focused on feeding its oil habit than kicking it," he wrote.[2] We couldn't agree more.

Let us be clear. There is nothing fundamentally wrong with domestic drilling, and in light of the economic and security threats to the nation, keeping 51 percent of America's oil and 27 percent of natural gas reserves off limits for exploration seems like undue deference to the vocal environmental lobby. More oil production in the United States means fewer dollars leaving our economy in exchange for foreign oil. As it is, net imports of petroleum in 2008 accounted for nearly half of the U.S. trade deficit in goods. If this money were to go to Exxon instead of Saudi Aramco or Hugo Chavez, more money would be ploughed back into the U.S. economy as opposed to the Saudi or Venezuelan economies. As much as Americans love to hate Exxon, it still pays corporate taxes in the United States, submits quarterly reports to its shareholders and adheres to strict environmental, labor and anti corruption practices. Can we say the same about Saudi Aramco and the other national oil companies?

But beyond the economic benefit, drilling will do almost nothing to address the U.S. strategic vulnerability. Drill anywhere you want, the United States has merely three percent of the world's conventional oil reserve while consuming roughly a quarter of the world's supply. The same figures are true for natural gas. If we continue to play in the petroleum playing field, we will forever be dwarfed by the Arabs, Nigerians, Russians and Venezuelans. Contrary to Gingrich's promises, if we drill here and drill now we will not pay less, at least not in the foreseeable future. A May 2008 analysis by the Energy Information Administration estimated that under the best-case scenario opening up ANWR would reduce prices by $0.41-$1.44 per barrel by 2027.[3] Drilling off the continental U.S wouldn't affect prices until 2030.

More importantly, experience of the past three decades clearly shows that whenever non-OPEC producers like the U.S. increase their production, OPEC decreases supply accordingly, keeping the overall amount of oil in the market the same. In other words, when we drill more, OPEC drills less.

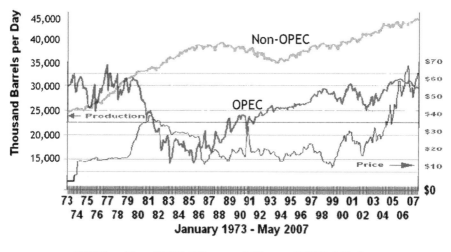

OPEC vs. Non-OPEC. When we drill more OPEC drills less
Source: WTRG Economics

Furthermore, the problem with the national discussion on domestic drilling is that it has been framed in pure environmental terms (are you a friend or an enemy of the caribou?) when what is really at stake is not the fate of a remote nature reserve but the ability of future generations of Americans to enjoy a modern lifestyle throughout the 21st century and beyond. While today most of our oil is used for ground transportation, a sector in which one can certainly foresee significant oil displacement, there are many other uses for oil which are just as important but where oil cannot be as easily displaced, including pharmaceuticals, petrochemicals, aviation fuel, detergents and plastics. Therefore, a nation that consumes a quarter of the world's oil but owns only three percent of conventional world reserves should debate the drilling issue in the context of one critical question: How much oil do we want to leave for future generations so they too can paint their homes, pave their roads with asphalt, make their medicines, drink from plastic bottles, wear rubber-soled sneakers, brush their teeth and enjoy multiple other applications of petroleum? In the next three decades, the U.S. population is projected to grow by 60 million. Assuming that by 2030 we succeed in eliminating the need for oil in the ground transportation sector, we will still need every single year, for other purposes, an amount of oil equivalent to more than five percent of all the technically recoverable oil contained in ANWR and offshore combined. But for the "drill here, drill now, pay less" cohort, America's future needs for petroleum take lower priority than short-term political gains. This is hardly a surprise when the next generation of Americans can expect to be greeted with a $10 trillion national debt and a crumbling Social Security system. Compare this with Saudi Arabia, where per capita oil reserves are 130 times that of the U.S. Upon recent discoveries of more oil in the kingdom, King Abdullah ordered that those new finds be left untapped to preserve the nation's oil wealth for future generations. "When there were new finds, I told them, 'No, leave it in the ground, with grace from God, our children need it,'" the king said.[4] Behind the king's statement lies a plain truth: The Saudis prefer to sit on their oil, while we are rushing to deplete ours. The Saudi reserve-to-production ratio – an indicator of how long proven reserves are likely to last at current production rates – is 70 years; Iran's is 82; the United Arab Emirates' is 90; and Venezuela's is 91. Iraq and Kuwait are at more than 100. How long does the U.S. have left? Eleven years. While America is increasingly dependent on foreign oil, to the detriment of its national and economic security, making domestic drilling the linchpin of our energy policy will perhaps buy us a few more years of complacency in the driver's seat of our SUVs, after which we will be guaranteed to be in a world in which almost no oil can be found outside the Middle East.

Efficiency diet

While conservatives on the right are in love with drilling, in the universe of liberals, moralists and enviro-visionaries, the idea that we can solve our energy dependence problem by simply consuming less has become sacred scripture. In a society where people live in the most spacious homes, drive the largest cars and eat the biggest portions, it is understandable why there is a strong sense that there is much to cut from. And indeed there is. Efficiency is no longer a "personal virtue" as former vice president Dick Cheney once described it. It is our generational responsibility if we want to accommodate the needs of 2.4 billion Chinese and Indians that are now joining the consumerist party with hundreds of millions of toaster ovens, television sets and cars without blowing the world up in a series of ugly resource wars. Simply put, we have to use less in order for more of us to live. The good news is that it works. The last time the United States made a concerted effort to improve energy efficiency – between 1979 and 1985 in response to OPEC's oil embargo – fuel efficiency of the average American car doubled and U.S. oil consumption decreased by 15 percent. Sadly, since the 1980's fuel efficiency has been on the decline. Americans seem to favor large trucks and sports utility vehicles, and fuel efficiency lies low on the priority list of the average American driver.

But there are some who never gave up on efficiency, even when it was unpopular. The most renowned of them is the indefatigable energy-efficiency guru Amory Lovins who two decades ago co-founded the Rocky Mountain Institute (RMI) in the mountains of Snowmass, Colorado. From this efficiency laboratory came some of the most interesting ideas on how to make more with less. In a Pentagon sponsored study *Winning the Oil Endgame: Innovation for Profits, Jobs and Security,* Lovins and his team put together a compelling case for market-oriented and innovation-driven solutions requiring no taxes and no mandates. His theme is simple: if we do the right things, it will cost less to displace a barrel of oil than to buy one. Lovins puts huge faith in advanced composite or lightweight materials that can nearly double the efficiency of today's cars while improving safety and performance. The RMI team claims that the vehicle's total extra cost is repaid from fuel savings in about three years. In addition to a national program to scrap clunkers, they propose government "feebates" aimed at shifting customer choice by imposing fees on inefficient vehicles while offering rebates to those who buy efficient vehicles. On the manufacturing side, Lovins called for $70 billion of taxpayer funds to support U.S. automakers in their effort to make advanced technology vehicles. All together, he calculates it would take a $180 billion investment over a decade to yield a $130 billion *annual* savings by 2025.

Had Detroit listened to the RMI's suggestion that these investments should earn a handsome return, with big spin-off benefits, when the report came out in 2004, it may have had a chance to avoid ending up in the financial dire straits in which it found itself four years later. But at that time, gasoline prices were low, fuel efficiency was not a major consumer priority, and the Big Three were focused on selling SUVs and other gas guzzlers rather than fuel efficient cars. For nearly three decades, the auto industry effectively blocked efforts to increase fuel efficiency standards. At the same time, the huge tax write-offs allowed for small businesses buying vehicles weighing over 6,000 pounds – the largest of the large SUVs – served to encourage their purchase. With rising oil prices and the Democrats taking over Congress in November 2006, this opposition was finally broken. In December 2007, Congress and the Bush Administration enacted the Energy Independence and Security Act of 2007, a bill that for the first time in more than two decades increased mandatory fuel efficiency standards, lifting the required fleet average efficiency to 35 mpg by 2020. In May 2009, the Obama administration announced an even more aggressive increase in efficiency: 35.5 mpg by 2016.

Such a leap in efficiency, we are told, could push oil use down by about 2.5 million barrels per day. While the new standards were widely applauded, many efficiency advocates believe that much more can be done. The Union of Concerned Scientists, for example, holds that our fleet can reach an average of about 40 mpg by 2020 with conventional technology alone, and more than 50 mpg by 2030 with the additional deployment of hybrid-electric technology.[5] But today, Detroit is effectively on its deathbed, and it is not even clear that it will be able to comply with the new mandate, not the least with a far more stringent one.

In and of itself, increasing efficiency is a worthy effort. If you can travel more miles for the same amount of money – why not? To a degree, this is already happening. The automakers have begun to work toward better fuel efficiency through improved transmissions, reduced weight and hybrid technology. Automakers, at least those that will survive the recession, are already on their way to offer hybrid versions of all their car models. Yet, as with drilling, efficiency too does not address the main enabler of oil's monopoly in the transportation sector – the gasoline only car. More efficient cars, even those the Union of Concerned Scientists would like to see on the road, certainly use less oil, but they still use nothing *but* oil. Therefore, they do not help us achieve energy independence. Boosting the fleet-wide gas mileage to 35 mpg in 15 years is like forcing a 2 pack a day smoker into a 15-year withdrawal plan after which he or she will smoke only a pack and a quarter a day. The question is: will the patient last that long? In this case, most doctors would probably consider a much more aggressive approach, which could show results in time for the patient to survive.

Furthermore, just like in the case of drilling, efficiency alone leaves the power in the hands of OPEC. In 2008, due to the meteoric rise in gasoline prices and the subsequent economic slowdown, U.S. gasoline demand dropped by nearly 10 percent. This was as if the U.S. fleet increased fuel efficiency by 2.5 mpg overnight. Improving fuel economy by that much would save the United States almost one million barrels per day. What was OPEC's response? First, it called for an emergency meeting in Vienna in October where it reduced production by 1.5 mbd. A following meeting in Cairo in November reaffirmed the decision. Then, in December that year, OPEC members met again to further reduce production by additional 2.2 mbd. Altogether OPEC dropped production by roughly 4 mbd since the economic slowdown of 2008 began, which is more than the amount of oil that was actually saved due to reduced consumer demand. Strategically, domestic drilling and increased fuel efficiency are two sides of the same coin: when non-OPEC countries drill more, OPEC drills less, and when we use less, OPEC also drills less.

Efficiency is important. It reduces waste of non-renewable organic molecules that nature labored millions of years to create. It helps reduce our trade deficit at a time our national debt is already in the stratosphere. It reduces the amount of pollutants our cars release to the detriment of our health. But let's not kid ourselves: it will not break our dependence on oil. Or as our colleague Robert Zubrin, author of *Energy Victory* put it: "Conservation through gasoline efficiency is [...] like trying to survive in a gas chamber by holding your breath. We need to break out of the gas chamber." [6]

The same is true for another demand constraining policy that in the wake of the fall in oil prices has garnered support by a growing coalition of pundits – the gasoline tax. Or more precisely, an increase in the federal gas tax with a decrease in payroll taxes. "A gas tax reduces gasoline demand and keeps dollars in America, dries up funding for terrorists and reduces the clout of Iran and Russia at a time when Obama will be looking for greater leverage against petro-dictatorships. It reduces our current account deficit, which strengthens the dollar. It reduces U.S. carbon emissions driving climate change, which means more global respect for America. And it increases the incentives for U.S. innovation on clean cars and clean-tech. Which one of these things wouldn't we want?" Thomas Friedman wrote in December 2008.[7] The following month, conservative columnist Charles Krauthammer published an article in the *Weekly Standard* contending that a gasoline tax with a payroll offset is "once-in-a-generation chance" for America.[8] Here is how the "Net-Zero Gas Tax" works: "A $1 increase in the federal gasoline tax - together with an immediate $14 a week reduction of the FICA tax. [...] The math is simple. The average American buys roughly 14 gallons of gasoline a week. The $1 gas tax takes $14 out of his pocket. The reduction in payroll tax puts it right back." There is great appeal for this proposal. No money is taken out of the economy so Washington doesn't get fatter. Nor does it get leaner. "It is simply a transfer agent moving money from one activity (gasoline purchasing) to another (employment) with zero net revenue for the government."[9] It's an elegant way to become more fuel efficient without forcing Detroit to manufacture small, unattractive cars, Krauthammer claims. And he is right. This policy would indeed force us to use less gasoline, leaving consumers to decide whether to drive less or shift to a more efficient car. The bad news is that – as long as most cars can be powered only by gasoline or diesel – the reduction in oil demand driven by a higher tax on petroleum fuels wouldn't make a strategic difference: just like in the case of more domestic drilling or higher CAFE standards, OPEC would react simply by drilling less. A higher gas tax in and of itself wouldn't alter our fundamental energy vulnerability – oil's virtual monopoly over our transportation fuel. So long as vehicles run solely on oil, no matter how one rearranges the market, OPEC wins and we lose.

The Cheney plan to export the Middle East morass

Dick Cheney was one of the most controversial figures in American politics and arguably the most influential Vice President in recent history. His fingerprints were found on almost every key policy initiative during the eight years of the Bush Administration from the Iraq War to NATO expansion. As a former oilman, he certainly had a great deal of impact on U.S. energy policy. Soon after President George W. Bush took office, Cheney released a national energy policy report by a group of undisclosed industry insiders, officially known as the National Energy Policy Development Group (NEPDG). The report drew a great deal of fire not so much due to its content but rather because of the secretive manner in which it was conceived and authored. Until its last day in office, the Bush Administration refused to reveal the names of the task force's participants, though in July 2007, the *Washington Post* reported the names, most of them from energy-producing industries. In addition to calling to increase domestic energy production, the main policy thrust of the Cheney plan was diversification of U.S. oil imports and the strengthening of global alliances to allow that. It advocated the reduction of U.S. dependence on the Middle East through increased reliance on the new oil El Dorados that rose since the 1990s in places like West Africa, Latin America and the Caspian basin. The plan called for the U.S. government to "deepen bilateral and multilateral engagement to promote a more receptive environment for U.S. oil and gas trade, investment, and operations," to "promote geographic diversification of energy supplies," to "improve dialogue among energy producers and consumers" and to "conduct a comprehensive review of sanctions" to allow U.S. oil companies to drill for oil in previously prohibited sanctioned oil domains like Iraq and Libya. In other words, the Cheney plan was in essence an articulation of the intent to continue and expand long standing policy by both Democrat and Republican administrations of subjugating U.S. foreign policy to America's energy needs. This was a natural extension of a policy first clearly articulated by President Carter in 1980 that became known as the Carter Doctrine, and in essence said that the United States is prepared to use all the means at its disposal, including military force, to defend access to oil.[10]

Since the 1990s, countries like Kazakhstan, Azerbaijan, Nigeria, Angola, Equatorial Guinea and Colombia became centers of a flurry of U.S. diplomatic and military activity. The United States invested billions in building bases, selling weapons and training local armed forces to protect oilfields, pipelines and refineries. In Central Asia, the United States has developed close military ties with oil rich Azerbaijan and Kazakhstan and until its eviction in 2005 and 2009 respectively, it used air bases in Uzbekistan and Kyrgyzstan to support military operations in Afghanistan. Georgia has no oil, but it is the pivot of the Baku-Tibilisi-Ceyhan pipeline, the main conduit of Caspian oil to Western markets. This merited U.S. military aid to the Georgian military and steadfast support for the Georgian government before and during its war with Russia in the summer of 2008. In West Africa, "one of the fastest growing sources of oil and gas for the American market," according to the Cheney report, U.S. military assistance has been growing by leaps and bounds, and in 2008 the U.S. military even established a new command, the African Command (AFRICOM) to defend U.S. interests in the region. Not surprisingly, the principal military aid recipients in the region are Angola and Nigeria, the two largest oil exporters. The Pentagon is currently seeking to acquire a base in oil rich Sao Tome, a tiny island state off the coast of Nigeria, where only 190,000 people live. In Colombia, where significant untapped reservoirs are believed to exist, U.S. Special Forces helped protect oil pipelines belonging to Los Angeles based Occidental Petroleum from guerilla attacks. Never in history has the U.S. military been so preoccupied with the mission of providing oil-protection services as it is today, and that without even mentioning the incredible, and expensive, presence of U.S. forces in the Persian Gulf.

At first glance, diversification of sources may seem to be a sound approach. But this solution is no more than a Band-Aid, and, in the long run, could breed stronger reliance on the club of countries on which the United States would like to be less dependent. There are three downsides to the Cheney approach. First, oil is a globally traded, fungible, commodity, so stifling U.S. purchases from the Persian Gulf and buying from other regions like Africa would just mean that somebody else would buy more from the Persian Gulf with no impact on price and availability, certainly not on oil's status as a strategic commodity. Think of the swimming pool analogy discussed earlier. Second, reserves outside of the Middle East are being depleted almost twice as fast as those in the Middle East. The overall reserves-to-production ratio in non-OPEC countries is about 15 years comparing to roughly 80 years in OPEC. With the growth in global demand, many of today's large, non-Middle East producers such as Russia, Mexico, Norway and China are running a marathon at the pace of a sprint, and, if production continues at today's rate, many of today's largest producers will cease to be relevant players in the oil market in less than two decades. At that point, the Middle East will be the remaining major reservoir of abundant, cheap crude oil and the world's dependence on it will grow rather than diminish. This could allow Middle Eastern producers even more leeway than they have today to manipulate prices and increase their political leverage on U.S. foreign policy. Third, deepening alliances with various non-democratic African and Central Asian energy exporters undermines America's foreign policy priorities like human rights and democracy promotion. Supplying non-democratic oil producers with advice and state-of-the-art weapons enables these regimes to stay in power and oppress their people with impunity. As mentioned before, such relations have proven in the past to be extremely problematic and in conflict with America's prime foreign policy goal of spreading freedom and democracy in countries where they are in deficit. Just like the Middle East, both Central Asia and West Africa suffer from territorial disputes, authoritarian regimes, bad governance, corruption, ethnic and religious strife and terrible human rights records. Nigeria, expected to supply a quarter of U.S. oil imports by 2030, is one of the most corrupt countries in the world and despite its oil riches most of its people live on less than $2 a day. The situation in Angola, Chad and Equatorial Guinea is not much better. Central Asia's most important producers, Azerbaijan and Kazakhstan, both have human rights records that would normally deny them U.S. support. Kazakh ruler Nusultan Nazarbayev's constitutional change allowing him to be president for life didn't generate a squeak from Washington. By becoming increasingly dependent on new energy producing regions, the United States is forced to turn a blind eye to their social ills and in doing so it undermines the prospects for the kind of reforms that are the keystone of its own diplomatic efforts. Such new dependencies open the door to kowtowing to dictatorial regimes and in essence help export the Middle East morass to the rest of the world. Cheney's plan might improve America's energy security in the short run, but in the longer run it would only replay in other arenas the problems the United States currently faces in the Middle East.

The Pickens Plan: A Boon for Pickens, not for America

How many times have you heard politicians and pundits claiming that to reduce oil dependence we should invest in nuclear, solar or wind power? Claims about nuclear power's potential role are often made by conservatives like Bill O'Reilly, Charles Krauthammer and Newt Gingrich. In his 2008 presidential campaign, Senator John McCain's most highlighted solution for America's oil dependence was to build 45 nuclear power plants. Solar and wind are the Left's favorites, appearing numerous times in Barack Obama's presidential campaign speeches. No surprise then that an NBC/Wall Street Journal poll from June 2008 showed that, when asked what would be the most effective policy in addressing the energy crisis, 37 percent thought that it would be the development of solar, wind and nuclear power. These protestations had merit more than three decades ago, when a significant part of our electricity was generated from oil or today, if you happen to live in Hawaii, the only state in the union where most electricity is petroleum powered. But in the rest of the 49 states, we essentially no longer generate power from oil. Not even in Alaska. To be precise, only two percent of U.S. electricity is petroleum based and only two percent of U.S. oil is used to generate power. So build all the solar panels, wind farms and nuclear reactors you want, and all that electricity would only compete against the electricity supplied today by our coal and natural gas stations. It will not displace oil. Electricity and transportation are two completely separate issues. Increasing the role of solar, wind and nuclear power, while it would have some environmental benefits, will do nothing to address oil dependence, at least until electric cars enter the market in significant numbers, though even today, cars plugging into the grid to charge their batteries are plugging in to non-petroleum based power.

If this is so evident, then why are so many confused? For many who still remember President Carter in his cardigan sweater calling on us to turn down the thermostat in our homes and install solar panels on our roofs, the story about how utilities nationwide weaned themselves from oil was somehow lost in the shuffle. Add to that a sustained campaign by some environmental groups to associate wind turbines and solar panels with our oil predicament and what we get is an urban myth of huge proportions.

And just in case the public wasn't confused enough, in July 2008 came Texas oil billionaire T. Boone Pickens with a plan to end our dependence on foreign oil based on a sophisticated way to force wind power into the oil debate. Trumpeted by a $58 million multimedia campaign, the Pickens Plan claimed to dramatically reduce oil use by shifting the transportation sector from petroleum-powered cars and trucks to natural-gas-powered vehicles. This would allegedly reduce oil imports by more than 30 percent and would supposedly save the U.S. economy $300 billion that otherwise would end up in the coffers of oil-rich foreign countries. According to the plan, wind energy would substitute for natural gas, now generating 15 percent of the nation's electricity, freeing natural gas to power a third of the vehicles in the United States. Pickens put his money where is mouth is (though some think he put his mouth where his money is) leasing hundreds of thousands of acres for a giant wind farm in the Texas Panhandle, where he plans to erect 2,700 turbines, but the plan he calls "a bridge to the future" is in fact a bridge to nowhere which could do more damage than good to U.S. energy security. Pickens' assertion that increased use of wind power would displace natural gas is based on wishful thinking. Our energy system is not a game of Lego – one piece can't replace another at whim. Even if 78 other billionaires were willing to follow Pickens' footsteps and build a 4,000-megawatt wind farm – that's the number needed to displace the current electricity production from natural gas – there's no way to guarantee that natural gas will be the primary electricity source displaced by all those turbines. Why not coal, or nuclear, or solar? Furthermore, implementation of the Pickens Plan would actually tie more natural gas to the power sector. Wind is an intermittent source of power – the wind doesn't blow 24 hours a day, seven days a week – and until and unless our electricity grid has sufficient power storage capacity, utilities counting on wind need to have backup power plants that can be powered up to fill in the gaps when the wind does not blow. This back-up power is today generally provided by natural gas.

There is even a bigger flaw in Pickens plan. Pickens claims that a shift from oil to natural gas would strengthen U.S. national security. His fascination with natural gas stems from the fact that 98 percent of the natural gas used in the United States is from North America, compared to more than 60 percent of our oil that is purchased from foreign nations. In other words, shifting from oil to natural gas means shifting from an imported resource to a domestic one. The problem is that in relation to its needs, the United States is not rich in natural gas, certainly not in sufficient quantity to power our cars and trucks for decades to come. Just as with oil, the United States consumes roughly a quarter of the world's natural gas but has only three percent of the world's conventional reserves. Its reserve-to-production ratio is less than 10 years. A shift to natural gas could even weaken U.S. national security: most of the world's reserves are concentrated in five countries – Russia, Iran, Qatar, Saudi Arabia and the United Arab Emirates – countries that are already engaged steadily and stealthily in discussions on the establishment of an OPEC-like natural-gas cartel. The Group of Three alone – Russia, Iran and Qatar, also known as the "troika" – exercises control over close to two-thirds of the world's natural gas reserves and a quarter of all natural gas production. Not very different from OPEC which controls more than three-quarters of the world's oil reserves but only a third of global production. Shifting from dependence on one cartel to another is like jumping from the frying pan to the fire. It's also the best gift the United States can give Iran – the world's second biggest holder of natural gas – at a time when we should be working to weaken Tehran economically.

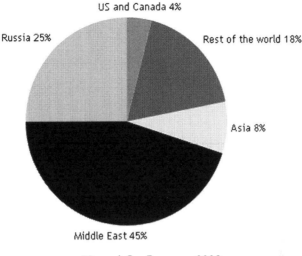

US and Canada 4%

Russia 25%

Rest of the world 18%

Asia 8%

Middle East 45%

Natural Gas Reserves, 2008

Sniffing hydrogen

The Pickens Plan is another reminder of how susceptible the public is to misguided and simplistic ideas that may sound good in theory but are devoid of scientific merit or are utterly unworkable. Perhaps the biggest of all energy hoaxes was the "hydrogen economy" – a vision of an economy based on millions of stationary and portable fuel cells fueled by pure hydrogen gas. Because hydrogen is the most abundant element in nature, many environmentalists and pundits were misled to believe that hydrogen could become a global energy 'currency' that could enable every country to meet its own energy needs through hydrogen production from renewable sources like solar power, wind turbines or hydroelectric power. In the ideal world of the hydrogen proponents, cars, trains, ships and planes would travel great distances carrying on board highly pressurized or liquefied hydrogen. Used in a fuel cell, hydrogen emits nothing but hot water vapor, and with worldwide use of "zero emissions vehicles" humanity could address its emissions problem. Enviro-visionaries like Jeremy Rifkin, Peter Hoffmann and Amory Lovins also foresaw a revolution in the marketing and distribution of electricity in the U.S. According to the vision, in a Hydrogen Economy, every fuel cell owner could potentially become an energy producer. With millions of end users, the monopoly of electricity generation would be taken away from utility companies and a worldwide "hydrogen energy web," to use Rifkin's term, similar to the World Wide

Web would be created. The outcome: a clean planet, every human being and community would be "empowered" by the ability to generate energy and the creation of "the first truly democratic energy regime in the world." This wonderful vision captured the imagination of journalists like *The Economist*'s Vijay Vaitheeswaran whose book *Power to the People: How the Coming Energy Revolution Will Transform an Industry, Change Our Lives, and Maybe Even Save the Planet* was nothing less than an ode to hydrogen. So infatuated was Vaitheeswaran with the hydrogen economy that his only criticism of President Bush's promise in his 2002 State of the Union Address that "a child born today will be driving a hydrogen, pollution-free vehicle as his or her first car," was that it was missing "the energetic banging on some tom toms."[11]

The Bush Administration's rally behind hydrogen was costly both in terms of taxpayer dollars as well as wasted years in the move away from oil. In a surprising January 2002 announcement, the U.S. Department of Energy declared the termination of the $1.5 billion, eight year joint project with the Big Three carmakers to develop high mileage cars and instead committed itself to the development of hydrogen fuel cell vehicles. Then Secretary of Energy Spencer Abraham coined the project "Freedom Car," and a linchpin of President Bush's National Energy Plan to reduce American reliance on foreign oil.[12] The federal effort whetted the states' appetites. California Governor Arnold Schwarzenegger quickly retrofitted one of his Hummers to run on hydrogen and declared the formation of a "hydrogen highway" connecting Sacramento and Los Angeles. In April 2002, Michigan Governor John Engler unveiled a plan called Next Energy which aimed to make the auto state the world leader in developing, manufacturing and marketing hydrogen technologies.[13] General Motors, for its part, announced that it would have a million hydrogen cars on the road by 2010. "The hydrogen economy is within sight." Summarized Jeremy Rifkin in his 2002 book *The Hydrogen Economy*, "For the first time in human history, we have within our grasp a ubiquitous form of energy, what proponents call the "forever fuel.""[14]

While to the untrained eye the hydrogen vision sounded plausible, many scientists looked at each other in utter disbelief in the face of what could only be described as yet another establishment-sponsored boondoggle. That hydrogen is not an energy source but an energy carrier that must be either extracted from primary energy sources like coal or natural gas, or extracted from water using another energy carrier, electricity, which happens to be also generated primarily from coal or natural gas, was a scientific fact our leaders neglected to note. Ask yourself this: what's more efficient:

1. using electricity to split water into hydrogen and oxygen, then storing that hydrogen on board the vehicle at high pressure (not easy since hydrogen is the lightest and thus leakiest gas) then using a fuel cell to convert that hydrogen back to electricity.
2. plugging the vehicle in and using the electricity directly.

Pretty obvious isn't it?

The other common method to generate hydrogen is to extract it from natural gas, in a process called reforming. In the above section about the Pickens Plan, we discuss why wedding our transportation sector to natural gas wouldn't be an improvement from an energy security perspective. Our leaders also didn't give sufficient weight to the multiple unanswered technological challenges related to hydrogen safety, packaging, delivery, storage, durability and cost. How could the 175,000 service stations in the U.S. be retrofitted to serve hydrogen at a bank-breaking cost of $1.4 million a piece (if the hydrogen was extracted using renewable energy, the cost would jump even higher to $2.3 million) especially when safety requirements barred positioning hydrogen pumps near gasoline pumps? No worries, they thought, build the cars and the rest will would magically follow.

As we know today, the cars were never built beyond prototypes and the rest never followed. Five years and billions of dollars later, the hydrogen economy essentially reached a dead end. President Bush left the White House and his successor hasn't uttered one good word about hydrogen. GM failed to put even one hydrogen car in the showroom, and in any case filed for bankruptcy, and the California Terminator is now busy convincing Californians that electric cars and plug-in hybrids are the wave of the future. Vijay Vaitheeswaran whose techno-optimism regarding hydrogen infected so many, went on to write another bestseller *ZOOM: The Global Race to Fuel the Car of the Future*, which was far less bullish about the technology. Governor Engler moved on to Washington to be the President of the National Association of Manufacturers, the nation's largest trade association while the State of Michigan leads the nation in unemployment. Spencer Abraham became a consultant to Persian Gulf governments and, in 2007, the ambassador to "official Washington" in the Fred Thompson presidential campaign. Finally, in May 2009, the Obama administration, after reaching the conclusion that hydrogen fuel cell technology will not be practical over the next 10 to 20 years, decided to cut off funds for the program.[15]

How an entire nation with its pundits, politicians and industry leaders fell for the hydrogen hoax is a topic for another book if not a national commission of inquiry. Such erratic decision-making processes, the failure of sloppy journalists and scientifically illiterate politicians to ask the right questions and the tendency to hype and prematurely pick R&D winners is symptomatic of our energy policy. Perhaps if we had more *real* scientists and fewer *political* scientists in "official Washington" and its think tanks, such colossal mistakes would have been avoided. But with fewer than a dozen of the 535 senators and congressmen representing us in Washington having a scientific education, even pork can fly.

⌘   ⌘   ⌘

4
FROM THE CAR AHMADINEJAD LOVES
TO THE ONE BIN LADEN HATES

What do you call a world leader who faces a strategic threat stemming from his country's vulnerability to gasoline supply disruption and, to address the problem, decides to introduce a crash program for energy independence that taps into his country's domestic resources and uses existing technology? Ahmadinejad. Iranian President Mahmoud Ahmadinejad is fully aware of his country's Achilles Heel: its lack of refining capacity to meet domestic demand for gasoline and other essential refined petroleum products. With more than 40 percent of Iran's gasoline imported, the Iranian leader knew that a comprehensive gasoline embargo cause social unrest that could undermine his rule. He is therefore took action to protect his regime. In 2005, he announced a three-prong program for 'energy independence' aimed at reducing Iran's dependence on foreign gasoline. One tenet of the plan is massive expansion of the country's refining capacity. Iran is investing $23 billion to build seven refineries and related plants. A second pillar of Iran's plan is to secure imports of refined products from Venezuela, one of its staunchest allies against the West. The third, and most innovative, part of the plan involves converting Iran's vehicles to run on compressed natural gas rather than gasoline. Unlike the United States, which is poor in conventional natural gas relative to its potential need, Iran has the world's second largest natural gas reserve after Russia. Its 920 trillion cubic feet of proven natural gas reserves, 16 percent of the world's total, essentially guarantee an uninterrupted supply of cheap transportation fuel for many decades to come. "If we can change our automobiles' fuel from gasoline to [natural] gas during the next three-four years," Ahmadinejad said in July 2007, "we won't need gasoline anymore." According to the Iranian plan, within five years most of Iran's passenger vehicles will be retrofitted to run on compressed natural gas and all of the nation's refueling stations will be converted to serve the fuel. The conversion of cars from gasoline to natural gas is simple, particularly in a country where unemployment is high and labor is cheap. All that is needed is a minor engine adaptation and the installation of a gas cylinder in the trunk of the car. Such bi-fuel cars can run on either gasoline or compressed natural gas or shift back and forth between the two. More than 105 conversion centers were built throughout Iran to which Iranians can drive in their gasoline only cars and several hours later pick up their cars, now able to run on cheap and clean burning natural gas. The cost of conversion of both the cars and the refueling stations – a sizable expense in both cases – is heavily subsidized by the Iranian regime.[1] For Iran, the shift from petroleum to natural gas has environmental as well as economic benefits. It will reduce urban air pollution and save Iran between $3 and $8 billion per year on gasoline imports. But for Ahmadinejad, the rationale for the plan is purely strategic. With most of Iran's cars running on natural gas, his refineries will be free to produce a greater proportion of essential non-gasoline petroleum products like jet fuel, which will keep his air force and commercial airlines intact, and diesel, to power his army and navy. Ahmadinejad's gas revolution is a clear sign that Iran is preparing itself for the possibility of war and is developing a comprehensive economic warfare strategy to supplement its military and diplomatic initiatives. This plan for "energy independence," which has largely gone unnoticed by the West, means that within a couple years Iran could be virtually immune to international sanctions.

The use of an innovative domestic energy policy to advance strategic goals is something we should all learn from Ahmadinejad. Fanatic and dangerous as he is, he understands what energy independence truly means better than most of his rivals in the West. His plan breaks the monopoly of gasoline in Iran's transportation sector by introducing competition with another transportation fuel with which the Islamic Republic is well endowed. He is not alone in doing so. In April 2009, another enemy of the United States, the firebrand president of Venezuela, Hugo Chavez, adopted a similar plan. According to his energy plan, at least 30 percent of the vehicles sold in Venezuela must be natural gas capable. Toyota, Volkswagen, Nissan, Chrysler and other automakers will be selling between 80 and 90 dual-fuel car models that can accommodate both gasoline and natural gas, providing drivers with fuel choice at the pump. According to the Natural Gas Vehicle Coalition, Venezuela's national oil company Petroleos de Venezuela plans to complete the installation of 350 new natural gas fueling stations this year, while 126 conversion centers will allow Venezuelan drivers to convert their gasoline-only cars to also run on natural gas. "A person who has a catanare [an old car that burns gasoline] - I know the solution: Give me your catanare and I give you a new car. However, this car will only use [natural] gas, won't use processed gasoline" Chavez said.[2]

Wonder why an oil rich country is investing in dual use vehicles? The reason is that gasoline in Venezuela is heavily subsidized and any attempt to lift the subsidy would be highly unpopular and potentially destabilizing. Chavez and Ahmadinejad have much in common. They are both defiant champions of anti-Americanism; they both have emerged as destabilizing forces in their regions; and they both represent the hard-line wing within OPEC, the one that is open about pushing for deep production cuts in order to push oil prices up. Despite their notoriety, there is one front on which America should take a cue from them: Chavez and Ahmadinejad are enabling in their countries something Americans don't have, but should – fuel choice. Indeed, despite their pomp, bluster and oil wealth, it is Chavez and Ahmadinejad, much more so than Obama, or any American president for that matter, who are currently leading the charge to break oil's monopoly in the transportation sector. While Obama's energy plan has put its sights on carbon dioxide emissions as the main energy challenge, Chavez and Ahmadinejad are holding the bull by its horns, implementing an energy policy that targets the actual use of oil based transportation fuel.

While Iran and Venezuela are moving quickly toward energy independence, Brazil is already there. It's a striking turnaround; three decades ago, the country imported 80 percent of its oil supply. Today, it is energy secure thanks to two steps it took. First, it eliminated its oil imports by drilling for oil offshore. But what truly made the country energy independent was the fuel flexibility of its fleet. Since the 1973 Arab Oil Embargo, the Brazilians have invested massively in their sugar-based alcohol industry. With warm temperatures and a long rainy season, Brazil has the perfect climate for sugarcane production. Since 2004, Brazil has shifted to a fleet of flexible fuel vehicles that can run on any combination of gasoline and alcohol. (As we shall soon see, alcohol does not mean just ethanol, and ethanol does not mean just corn.) The technology is a century old. Henry Ford's Model T was the first commercial flex fuel vehicle. The engine was capable of running on gasoline or ethanol, or a mix of both. Flex fuel vehicles differ from bi-fuel vehicles, such as Iran's natural gas vehicles, in which two fuels are stored in separate tanks and the engine runs on one fuel at a time. Flex fuel vehicles have one fuel tank, in which the different fuels mix. A flexible fuel vehicle is like a cup of coffee. Some like their coffee black, others with lots of milk. Some like cow's milk, others who suffer from lactose intolerance can use soy or rice milk, and a few may even prefer the more exotic coconut milk. The key is – you choose how much of which liquid goes into your cup. Similarly, a flex fuel engine allows you to choose how much alcohol versus gasoline goes into your tank. The car can handle almost any blend. (Almost. In cold climates, too much alcohol can cause engine starting problems. That's why in Brazil, where the climate is hot, cars can run on up to 100 percent alcohol, while in the United States, in colder parts of the country, 85 percent is the maximum level of alcohol in the blend, and in the even colder Sweden, the maximum is 75 percent.) In 2008, 80 percent of the new cars sold in Brazil were flexible-fuel vehicles. These cars cost less than $100 more to manufacture as compared to gasoline only cars. All it takes to turn a regular car into a fuel flexible one is a fuel sensor and a corrosion-resistant fuel line, since alcohol is more corrosive than gasoline. Lest anyone think that can't be done in the United States, many of the flex fuel vehicles sold in Brazil are made by General Motors and Ford.

What Brazil's flex fuel program did was open the once petroleum dominated transportation fuel market to competition. With the majority of their cars flex fuel, Brazilians can now choose between gasoline and alcohol at the pump. While between 2005 and 2008 fuel prices nearly doubled elsewhere, in Brazil, they were almost frozen. When oil prices soared in 2008, ethanol became so popular in Brazil that gasoline became an alternative fuel, and the Brazilian government had to step forward and subsidize the national oil industry. It gets even better. Like in Iran, in Brazil natural gas is widely used for transportation. Ten percent of Brazil's auto fleet, more than 1.5 million cars, can run on natural gas. Most of these natural gas enabled cars are also flex fuel vehicles. Such tri-fuel cars can run on gasoline, alcohol *and* natural gas. All this didn't happen by chance, but through perseverance and dedicated leadership. The result is enviable.

Another big country that seems to be on its way to adopt the Brazilian model of fuel flexibility is China. Since 1993, the year China became a net oil importer, China's oil imports have grown by leaps and bounds. In 2007, China imported 47 percent of its oil, and in 2008 it passed Japan to become the world's second largest oil importing nation after the U.S. Until 2006, China invested in expanding its ethanol industry to become the world's third-largest ethanol producer behind Brazil and the United States. But with soaring food prices in 2008, the Chinese government hit the brakes and banned the use of grain for alcohol production. With no active farm lobby or Iowa caucuses pushing corn use, China decided to veer toward another alcohol – methanol, a high performance fuel that was used for decades by drivers at the Indy 500. While ethanol is generally made from agricultural products like corn, sugar cane, sugar beet and, if new technologies prove economic, cellulosic material like switch grass, wood chips and other agricultural and forest residue, methanol can be made from all of the above plus natural gas, coal, garbage, any organic material and, perhaps in the future, recycled carbon dioxide. Today, about 90 percent of the worldwide production of methanol is derived from methane, the main component of natural gas (if T. Boone Pickens really wants to see more natural gas used as transportation fuel, methanol is the cheapest and most efficient way to get there.) China's interest in methanol was sparked by the fact that it, like the United States, is rich in coal, which can be converted to methanol for about 50 cents per gallon.

In September 2006, eight leaders from China's coal-producing provinces provided a report to Chinese President Hu Jintao titled "Suggestion on Promoting Methanol Fuels to Replace Gasoline and Diesel Fuel." President Hu approved this "suggestion" and directed the National Development and Reform Commission (NDRC), the arm of the Chinese government for macroeconomic planning, to explore the use of methanol fuels. The NDRC saw the potential of methanol for China's energy security and directed the development of national methanol fuel blending standards. As a result, China is the world's largest methanol producer and consumer. It also leads the world in the use of methanol as an alternative transportation fuel, blending nearly one billion gallons of methanol in gasoline annually. China's automakers have already begun to certify their cars to run on methanol. As Greg Dolan from the Methanol Institute wrote: "The country's fastest growing independent automaker, Chery Automobile, has recently completed demonstration work on 20 methanol flexible-fuel vehicles – capable of operating on methanol or gasoline – now ready for full-scale production. Shanghai Maple Automobile has announced plans to build 2,000 methanol cars in 2008. Chang'an has introduced the methanol-fueled BenBen car. Greely Automotive has put its Haifeng methanol car into production. Shanghai-based Huapa Automotive has built a number of methanol fueled cars. Shanghai Automotive Industry Corporation, one of the big three automakers in China, is developing a number of methanol-fueled cars. In addition, a number of smaller companies are converting large numbers of cars to methanol operation."[3] In the United States, where methanol flex fuel vehicles originally got their start, not one new car model, including the E-85 flexible fuel vehicle, is warranted to run on methanol.

They can do it, so can we

The stories of Iran, Venezuela, China and Brazil prove two points. First, breaking the monopoly of oil in a country's transportation system is possible and relatively affordable provided the national will and necessary leadership exist. Second, the payoff is fantastic both economically and strategically. As a result of its energy independence, Brazil was one of the most economically resilient countries in the face of the 2008 oil crisis. Can the United States do the same? Absolutely. Just like Iran, Venezuela, China and Brazil are reducing their vulnerability to transportation fuel supply disruption by enabling their auto fleets to run on something other than petroleum, so too can the United States. If Iran – with a per capita GDP of $8,300, China with $5,400, Brazil with $9,700 and Venezuela with $14,000 – can afford such an insurance policy, surely the United States, with its $45,000 per capita GDP, can too. The key is for vehicles to become platforms on which fuels can compete. How do we achieve this? As a first step, every new car sold in the U.S. should be a flex fuel vehicle warranted to use gasoline, ethanol, and methanol, a feature that costs less than $100 extra per car. We have multiple standard features in our cars today. Seatbelts and airbags protect our lives when we get into a collision. Rear view mirrors increase our field of vision. Our car radios are even mandated by law to receive FM transmissions so we can stay informed in time of crisis. An Open Fuel Standard requiring that every car sold in America be flex fuel can protect our lives and our economy more than all of the above. President Obama pledged numerous times to pass a law that would mandate flex fuel engines in all automobiles in order to break the oil monopoly. Secretary of Interior Ken Salazar, while still in the Senate, was the lead sponsor of legislation that required new cars sold in the United States to offer fuel flexibility. Energy Secretary Steven Chu has also spoken on the merits of this policy. "It would only cost $100 out of $15,000. Wouldn't it be nice to put in those fuel lines and gaskets so that we can use any ratio [of gasoline and alcohol] we wanted?," he said in June 2009.[4] Legislation that would ensure that most new cars sold in the United States by 2012 be flex fuel has been pending before Congress since 2005. In the congressional session of 2009-10, a bill called the Open Fuel Standard Act cosponsored by bipartisan teams of members is pending before both the House and the Senate. This bill would ensure that starting in 2012, 50 percent of new cars sold in the U.S. that have an internal combustion engine be warranted to operate on gasoline, ethanol, and methanol (and if they are diesel powered, warranted to use biodiesel), and starting in 2015, 80 percent.

U.S. automakers have been producing flex fuel cars for the American market for years, albeit in small quantities and only warranted for gasoline and ethanol. In 2009, there were no fewer than 33 car models, including the Dodge Caravan, Chevrolet Silverado, GMC Yukon and Ford Crown Victoria, warranted to run on up to 85 percent ethanol. Even the notorious Hummer H2 can now run on alcohol. Altogether there are close to 8 million flex fuel vehicles on U.S. roads today. That sounds like a lot of cars, but it is a very small portion of the overall vehicle fleet − not even 5 percent. That's not enough to make a dent in a fleet of more than 200 million vehicles. Outside of the Midwest, where corn ethanol is easily available, gas station owners for the most part aren't going through the hassle of retrofitting their pumps to serve alcohol (this despite large tax incentives which cover most of the cost), thus flex fuel vehicle owners are never able to exercise their option for fuel choice. Why? Put yourself in the shoes of the owner of a fuel station that has ten pumps and ask yourself what it would take for the business case to be there to convert one of those pumps to serve an alcohol fuel. The answer is, that until 15-20 percent of all the cars in your area are flex fuel vehicles capable of using that fuel, it doesn't make sense to allocate valuable retail space to such a pump, despite the tax credits that currently exist to cover part of the cost. The pump would just sit unused most of the time. With such low numbers of flex fuel vehicles as currently exist, the chicken and egg problem associated with the refueling infrastructure becomes insurmountable. The key step is to get to that 15-20 percent threshold. What is needed to get there is a far more significant penetration of flex fuel vehicles in the U.S., just like in Brazil. From the perspective of a vehicle owner, a flex fuel vehicle doesn't entail any inconvenience since if there are no alcohol pumps around it can be fueled with gasoline just as well. The way to solve the chicken and egg problem is to ensure that most new cars are flex fuel vehicles. With 10 million new cars rolling onto U.S. roads annually, a flex fuel mandate such as the one proposed by the Open Fuel Standard Act would ensure the critical 15-20 percent threshold is hit in just three years. At that point, fuel stations can start to catch up. U.S. automakers have repeatedly said they would be willing to make 50 percent of their cars flex fuel by 2012. All that is needed is for Congress to take them up on this commitment. Especially now, after billions of taxpayer dollars have been transferred by the federal government to auto company coffers with no end in sight, it seems like a minimal thing to get in return.

Lotus 270E: It's a flex fuel gasoline-ethanol-methanol vehicle
Source: Lotus Engineering

Why electricity?

In the next chapter we will discuss how to supply the fuel to power so many cars, but for now it is important to understand that critical as liquid fuel choice is, it is not sufficient to break the power of oil. Since the 1970s, OPEC has proved adept at manipulating oil supply to drop the price of oil whenever consuming countries appear to be making headway toward alternatives. While most liquid fuel options to gasoline are competitive around $45-50 a barrel oil, the constant volatility has often proved to give investors pause, stalling alternative fuel capacity expansion. The key to breaking this logjam is using electricity as a transportation fuel. It costs between a cent and a half and three cents a mile to fuel a car on electricity. For gasoline to be as cheap, oil prices would have to drop below ten dollars a barrel. This would be very difficult for OPEC to do and still keep OPEC countries' economies humming. If most cars on the road, in addition to being fueled by gasoline and a variety of alcohol fuels, were able to plug in to the power grid and be powered by electricity, then oil would face formidable competition. And since it would be very difficult for OPEC to increase supply sufficiently to drop prices lower than the ten dollars a barrel necessary to undercut electricity, then, as our fellow Set America Free Coalition member James Woolsey has pointed out, we can be pretty confident that they wouldn't bother to drop it just enough to undercut the alternative liquid fuels. That would be like dropping the price of Coke just enough to undercut Pepsi, but not enough to undercut Dr. Pepper, assuming they were all completely interchangeable.

Electricity is the fuel of the future. It's cheap, clean and domestically produced. Less than two percent of U.S. electricity is generated from oil. The infrastructure for electric transportation is already with us. We all have access to the grid, and, assuming you have a garage, all that is needed for your electric car to connect to the grid is an extension cord. Electric cars have been around in various configurations for more than a century. But for a variety of reasons they never quite made it. The main reason for their failure to capture the masses was their limited range. People want cars that offer unlimited range. No one wants to be stuck in the middle of a highway looking for a plug, or to wait six or eight hours at a rest station for the car battery to recharge. Most pure electric vehicles (EVs) offer a range of 80-100 miles and for many people this has been a non-starter despite the fact that half of the cars on America's roads are driven 25 miles per day or less.

Interestingly, 38 percent of U.S. households own two cars, and an additional 20 percent own three or more vehicles.[5] That makes over 64 million households in the United States with more than one vehicle which could conceivably replace one or more gasoline-only car with cars powered with made-in-America electricity. If each of those households keeps one liquid fuel car capable of making the long trip to visit granny, while buying an electric car for short haul commuting, a substantial part of our local miles could be traveled with electricity instead of gasoline. Environmental and national security concerns have created revived interest in EVs and GM, Nissan, Ford and Chrysler have said they will sell battery-only vehicles in the United States by 2011. The Tesla Roadster, a luxury all electric Lotus look-alike with a range of nearly 200 miles per charge, is already manufactured and distributed to those who can afford the cost – $109,000 per unit. In March 2009, Tesla displayed a prototype of its next car scheduled for delivery in 2011 – the Model S. This all-electric sedan which can seat seven people will have a range of 160 to 300 miles, depending on the chosen battery option at an estimated sticker price of

$57,000-$65,000. Despite the low cost of electricity compared to gasoline, the cost of this vehicle platform will still be prohibitive for most customers who expect to pay less than half that price for a family sedan. The range of a pure electric vehicle is a function of battery size and thus price. This means that until battery prices drop significantly, most customers will have to settle for EV's with a much smaller battery, which brings us back to the range problem.

One California based company, Project Better Place, has taken upon itself to extend the range of pure electric cars by providing an infrastructure for battery charging and replacement. The company, founded in 2007 by software executive Shai Agassi has signed agreements in Israel, Denmark, Portugal, California, Hawaii and Canada to erect charging stations for EV drivers in parking lots and other public spots. It also aims to introduce swapping stations that resemble car wash tunnels where batteries can be replaced in a matter of minutes. But regardless of the range, pure EV's have one major drawback.

Swapping one vulnerability with another

Electricity has huge potential as a transportation fuel. However, if all or even most of our cars were capable of running on *nothing but* electricity, our transportation system would be as susceptible to blackouts as our grid is today. This vulnerability is too often ignored by electricity-only enthusiasts. We do so at our peril. In the summer of 2003, a broken tree branch in Ohio started a cascading failure that brought to one of the worst power outages in history. Tens of millions of Americans remained without power for more than 24 hours. Massive as it was in scale, this blackout was relatively short term. But longer disruptions have also occurred. During Hurricane Rita in October 2005, hundreds of thousands of Houstonians fled the city causing vast traffic jams along the highways north of the city. Many motorists were stuck in their cars without gasoline in 100 degree heat.[6] A similar situation occurred in September 2008 when Hurricane Ike struck Houston leaving more than 2 million people in America's fourth largest city without power for more than a week. In both cases, Houston had no lights, no air conditioning and no hot meals. But its residents could still move around at will with their cars, taking care of their basic needs. One can only imagine the chaos had the exodus from Houston taken place on the wings of battery operated cars with an 80-mile range. Had most cars been able to run on nothing but electricity, the outcome could have been catastrophic.

The case of Hawaii is also worth highlighting. Only four weeks after the December 4, 2008 announcement by Governor Linda Lingle of her plan to team up with Project Better Place and electrify the cars on the Aloha State's roads, the island of Oahu lost power in the midst of heavy storm, leaving 800,000 residents and thousands of tourists in the dark. Hawaii is not as ideal a locale for EVs as is commonly believed. It is the only state in the United States that mostly uses oil for power generation. Considering that 77 percent of the state's electricity is generated from oil, Hawaii's demand for oil is more likely to grow as a result of the ambitious plan rather than diminish. But even if there were some hidden logic in shifting the Island's fleet from direct oil fueling to indirect oil fueling, the risk of putting all the eggs in one basket by putting the Island's auto fleet completely at the mercy of a centralized and vulnerable grid is not the most prudent thing to do. To be fair to the pure electric car, our gasoline refueling infrastructure is not exactly in top form during such disruptions as it too depends on electricity. But no doubt in such cases, uncommon as they may be, dependence on a limited range battery no longer seems appealing. Run all the cars on electricity and a single falling tree branch could cripple transportation for an entire region. Run them all on biofuels, and a drought could send fuel prices soaring. Run them all on gasoline, and a terrorist attack on a major oil facility in the Persian Gulf could bring our economy to a standstill. A resilient transportation sector is one that is open to the full range of fuel options, so no one disaster can bring it to a screeching halt. Cars that can run on a variety of liquid fuels in addition to electricity mean an insurance policy against natural or man-made disasters, maximum fuel choice for consumers and competition between socket and pump. That is exactly what flex fuel plug-in hybrid electric vehicles offer.

Plug-in Hybrid Electric Vehicles

Hybrid electric vehicles, such as the Toyota Prius, have an internal combustion engine and a liquid fuel tank. They also have a battery that, to simplify, gathers the energy that would otherwise be dissipated as heat during braking. Take any car – whether a big SUV or a small coupe – and make it a hybrid (don't add weight or power to the car, just hybridize it), and it will get about twice the miles per gallon it used to. While that's certainly good, it's not good enough: the hybrid's only external fuel source is still gasoline, and while it uses less than a conventional car, we still face the fundamental fact that when we use less, OPEC simply drills less. A no-plug hybrid doesn't address the core issue of breaking oil's transportation fuel monopoly. Take that same hybrid and soup it up – put in a larger battery and a plug – and you get a plug-in hybrid electric vehicle (PHEV.) Now the game is changed. Plug-in hybrids have smaller, lighter, less-expensive batteries than pure electric vehicles. But they have bigger batteries than conventional gas-electric hybrids, so they can run longer on batteries only. A PHEV with a 20-mile battery can run 20 miles on electricity, and when the charge is used up, automatically keep running on the liquid fuel in the tank, providing the standard three or four hundred mile range drivers are accustomed to. Simply put, it's like having a second fuel tank that you always use first – only you fill up at home, from an electric outlet, at an equivalent cost of under $1/gallon. PHEVs stretch each gallon of gasoline with electricity. In this way, PHEVs achieve oil economy levels of 100-150 miles per gallon of gasoline (mpgg) without compromising the size, safety, or power of a vehicle.[7]

PHEVs penetration into the public consciousness has been by no means smooth. At a time when the entire nation, its government, automakers and pundits were still consumed with the hydrogen mirage, expressing doubts about the government's pet project was borderline sacrilegious. But as hydrogen's drawbacks became apparent, "rebel groups" who quietly worked on more realistic solutions began to surface. Felix Kramer and a group of Menlo Park modern hot rodders from the California Cars Initiative (CalCars), Professor Andy Frank and his dedicated USC Irvine students, Robert Graham and Mark Duvall from the Electric Power Research Institute (EPRI), Dean Taylor from Southern California Edison, Roger Duncan with his Plug-in Partners Coalition and Chelsea Sexton, star of the movie *Who Killed the Electric Car* and founder of Plug in America, all played an important role in the PHEV revolution. So did we. On May 18, 2006, the CEOs of the Big Three automakers came to Capitol Hill to meet with congressional leaders and discuss the industry's woes. We saw an opportunity to introduce Congress to a solution that could make a big dent in our oil dependence and also help Detroit reinvent itself. After a whirlwind fundraising campaign, Felix Kramer air lifted his converted white Prius PHEV, built by EnergyCS with a lithium-ion battery, from California. His car was joined by a second Prius converted by Connecticut battery maker Electro Energy, this one equipped with a nickel-metal hydride battery. The two vehicles caused quite a stir in Washington, as everyone including dozens of Senators, Representatives, staffers, reporters, builders, tourists and even the Capitol police, took interest. This was the first time plug-in hybrids were seen in public in our nation's capital. The star of the show was the "dongle" – the short cable that plugs into the car's rear bumper outlet and connects to an extension cord for 120-volt power and comprises the car's entire "required infrastructure." That same afternoon, GM's CEO Rick Wagoner, Ford's Bill Ford and DaimlerChrysler's Tom Lasorda joined Michigan Representative John Dingell, the legendary chair of the House Energy and Commerce Committee, for a press conference outside the Rayburn Building. CalCars' John Davi who stood among the reporters asked the CEOs about the prospects for PHEVs. GM's Wagoner took the podium to explain how unviable PHEVs are, reiterating GM's commitment to manufacture hydrogen fuel cell cars instead. While he was describing the "insurmountable" problems facing PHEVs, the silver bullet with its "100mpg" signage made a triumphant drive behind his back, causing surprised reporters to turn their cameras from the CEOs to the new wonder. Six months later GM unveiled the Chevy Volt.

PHEV debut in Washington, May 17, 2006
Source: Set America Free Coalition

Beating GM to the race

If GM survives its financial woes, the Chevy Volt will be the first commercial PHEV produced by a major U.S. automaker. Rick Wagoner, who in March 2009 was sacked by President Obama as a condition of federal aid for the company, will not be in office to enjoy the glory. GM invested over $1 billion in this unusual car that can go 40 miles on a full plug-in charge. The Volt has a four-cylinder gasoline engine to recharge the batteries just like a generator. As a result, it potentially can use no gasoline for shorter drives and hit 50 to 100 miles per gallon on longer drives. Ford has also committed to sell PHEVs in the coming years. The company has already developed a PHEV version of its Escape that can travel about 30 miles on all electric power, reaching 120 mpgg in city driving. Toyota has gotten on the PHEV bandwagon as well, announcing that it, too, will unveil a plug-in hybrid prototype in 2010 using lithium-ion batteries. Jaguar, Hyundai, Volvo and Mazda have all made similar announcements. Legacy automakers aren't alone in the race for PHEVs; new players like Washington state-based AFS Trinity and California-based Fisker Automotive (founded by auto designer Henrik Fisker, whose creations range from the BMW Z8 to the Aston Martin DB9) are also developing their models. Fisker's Karma, a fancy four-door sports sedan PHEV with a 50-mile battery, is already being sold for $87,000.

An unknown Chinese company is beating all of the above in the high-stakes technology race. In 1995, Wang Chuanfu founded a battery company called BYD in Shenzhen, an industrial city in southern China just north of Hong Kong. Within a decade, the company emerged as the second-largest producer of rechargeable batteries to power electronic goods capturing 30 percent of the market for mobile phone batteries. In January 2003, Wang announced that BYD was acquiring a majority stake in a Chinese auto manufacturer in order for the company to become a player in the nascent electric car market. Investors ran for the exits, sending BYD's stock price tumbling. But Wang was not deterred. Against a wall of skepticism, he continued to develop the company's competence in auto manufacturing. In 2008, the skepticism faded away when the company surfaced on Warren Buffet's radar screen. Realizing its potential, Buffet bought a 10 percent stake in it. In December 2008, BYD dropped another bombshell when it presented the F3DM, the world's first mass produced PHEV. BYD took a full size family sedan costing $14,000, hybridized it, added a plug and a bigger battery that allows the car to drive 60 miles on a charge, shifting to its liquid fuel tank and internal combustion engine when the battery runs low. The retail price, $22,000, should terrify every Detroit executive.

The best of both worlds

So flex fuel vehicles or PHEVs? Alcohols or electricity? Which is a better solution to our energy challenge? Which should come first? Many energy independence advocates are so in love with one that they fail to see the benefits of the other. Sherry Boschert, a PHEV enthusiast and president of the San Francisco Electric Vehicle Association who wrote the thorough and otherwise excellent book *Plug-in Hybrids, the Cars that will Recharge America*, views flex fuel vehicles as "the latest distraction that delays introduction of plug-in hybrids and electric vehicles."[8] On the other side, some flex fuel vehicle champions see PHEVs as a diversion from *their* vision. Robert Zubrin for example pins his entire energy independence strategy on flex fuels, not mentioning PHEVs even once in his otherwise detailed and inspiring book *Energy Victory: Winning the War on Terror by Breaking Free of Oil*. Both camps are right in their advocacy and wrong on their attacks. PHEVs and FFVs are *both* essential pieces of the solution and in many ways they complement each other. They certainly don't undermine each other, and there is no reason to oversell or undersell either one. Add corrosion resistant fittings to the fuel line and tweak the software a bit, and for less than $100 extra, a PHEV can also be a flexible-fuel vehicle and drivers get the best of all worlds. A flex fuel plug-in hybrid is a car that runs on electricity *most* of the time and can shift to petroleum or non-petroleum liquid fuels *part* of the time. A PHEV would normally yield 100-150 miles per gallon of gasoline. If powered by say 80 percent alcohol and 20 percent gasoline, each gallon of gasoline is stretched with alcohol fuel by a factor of five, and oil economy could reach over 500 miles per gallon of gasoline. Notice we say 500 miles per gallon of gasoline, not 500 miles per gallon. The car isn't magically using a drastically reduced amount of energy – it's just stretching each gallon of gasoline with electricity and alternative liquid fuels.

The Energy Policy Act of 2005 (EPAct 2005), the first bill out of Congress to ever mention and provide for PHEVs, authorized a $40 million grant program to commercialize flex fuel plug-in hybrids – PHEVs that are also flex fuel. The idea was to advance the development of a truly flexible platform on which the broadest array of fuels could compete and which could achieve at least 250 miles per gallon of gasoline without compromising size, power, performance or safety. The sponsor of this section of the bill was none other than Senator Barack Obama. Unfortunately, no funds were ever appropriated for the program.

But Americans are not waiting for Congress. In March 2008, James Woolsey, former CIA director and the dean of the energy independence movement, collected the key to his converted, now plug-in, Toyota Prius. Woolsey's Prius was upgraded to a plug-in hybrid by Watertown, Massachusetts-based A123 Systems using an advanced Lithium-Ion battery which was developed at MIT. The conversion allows the Prius to achieve 100+ miles per gallon of gasoline and to travel up to 40 miles predominantly on electricity from one overnight charge. On the back of the car Woolsey has a bumper sticker "bin Laden Hates this Car," suggesting that if all of the cars were like this one, the jihadist movement would slide into bankruptcy. Woolsey, who lives on a farm south of Annapolis, Maryland, has solar panels on his roof so the battery of his converted Prius is powered by the sun. Each charge takes about 4 hours. "If my Prius had an ethanol, other alcohol, or bio-based diesel engine using 85 percent biofuels, I would get close to 500 mpgg on the petroleum fuel in the engine," Woolsey explained to President Bush who came to inspect the car. Across the country, in Los Angeles, another Prius owner was altering his car in a different way. Film director Jerry Zucker who is known for his hilarious productions like Airplane and Top Secret (check out his latest politically incorrect clips of a gas pump on the rampage at www.nozzlerage.com) turned his hybrid into a flex fuel car, using a $300 conversion kit sold by a Santa Monica-based company called White Lightening Ethanol Conversion Systems. The company offers to convert any gasoline-operated vehicle with a computer and fuel injection system to allow it to use both ethanol and gasoline. The conversion is typically completed in less than 45 minutes with a 5 year warranty. Woolsey's and Zucker's respective Priuses are in essence two halves of the same apple. It won't take long before someone will combine their two cars and produce a multi-fuel car that can run more than 500 miles on a gallon of gasoline. Such a car can provide consumers with the fuel choice they don't have today, allowing people in different states to decide daily on the energy mix that will carry them from one place to another. Whether you decide to run your car on Virginia coal, Iowa corn, Minnesota sugar beet, Brazilian sugar cane, Texas wind or Alaskan oil is up to you. The flex fuel PHEV, the car bin Laden hates, accepts them all.

⌘ ⌘ ⌘

5
FUELING 48 FLOORS

The United States consumes about 21 million barrels per day. What do 21 million barrels look like? If we attach 21 million barrels to each other we get a pipe 11,800 miles long, about twice the distance between New York and Beijing or half of the circumference of the earth at the equator. If we were to fill a container with 21 million barrels it would be roughly the size of the Twin Towers. Each one of the 110 floors of the towers would accommodate about 190,000 barrels, which is roughly equivalent to our daily import from Kuwait. The daily demand of the U.S. transportation sector fills the first 76 floors of the World Trade Center. The rest of the floors would supply the industrial, commercial and residential sectors. Within the transportation sector, the oil used to power our cars and trucks would fill up 48 floors. Those 48 floors are the crux of our energy independence effort. It's not that the others don't matter. Technology can certainly provide alternatives to oil in home heating, plastics and even aviation (albeit for the latter at quite a high cost, at least as of this writing.) But it is the virtual monopoly over transportation fuel in general, and ground transportation fuel in particular, which gives oil its strategic status.

As we explained in Chapter 2, to become energy independent we do not have to replace all the oil in all the 110 floors. We don't even need to displace all the oil in the 70 floors that simulate our petroleum imports or most of the oil in the 48 floors that make up the ground transportation sector. What we need is enough resilience in the system to ensure that the building remains standing when hit by a plane. The key step, as discussed in the previous chapter, is to ensure that vehicles enable fuel choice. To understand what impact fuel choice would have on the transportation fuel market, it is important to ask whether the market can create sufficient capacity in alternative fuels for real competition to be possible.

One of the fun parts of running an energy policy think tank is the daily interaction with some of the best and brightest scientists, inventors and entrepreneurs who dedicated their lives to the pursuit of alternative energy technologies. We have seen it all – from the conventional to the bizarre. Alternative fuel developers from landfill waste, algae, paper pulp, pig manure, turkey parts, restaurant oil, tires, onion skins, coffee grounds, auto dashboards, computer monitors, even solar energy from space, have all made their case to us in colorful brochures and detailed PowerPoint slides arguing that their technology is "the next big thing" and if Congress, the White House and Wall Street investors only learned about it (with our help of course) we could destroy oil once and for all. Many of the ideas making their way to our office are implausible. But every now and then there are also the lumps of coal that could, with some help and luck, turn into diamonds. Most liquid fuels are molecules of carbon, hydrogen and oxygen arranged in different orders and connected with different bonds, just like Lego pieces. Shuffle or change the ratios of those elements, break or create a bond or two and one liquid turns into another or into a gas or solid. Such conversion process may work well in a small demo but will it be economical on a commercial scale? The economics are what differentiates a scientist's field of dreams from a commercial success. Cost is critical, but it is not everything. Auto manufacturer acceptance is no less important in determining who can move to the marketplace and who stays on the drawing board. We have seen liquid fuels that could not make it to the fuel station simply because our cars were never warranted to run on them. Biodiesel, for example, is a domestically produced, renewable alternative to petroleum diesel fuel and can be made from plant oils, animal fats, recycled cooking oils or new sources such as algae. Until 2008, the biodiesel industry was in limbo because of a lack of commonly agreed upon approved finished blend specifications. The admirable efforts of home brewers supplying their own fuel notwithstanding, the absence of specs was the single greatest hurdle preventing automakers and engine manufacturers from including in their warranty the use of biodiesel blends in their diesel vehicles. While technologies to make biodiesel have been around for years, uniform biodiesel specifications were missing. These specifications were necessary to give automakers the confidence that biodiesel would have the same composition and thus the same behavior in cars every single time it was used, so they could test and warranty their vehicles for the fuel. This hurdle was finally removed in October 2008, when a rigorous process finally yielded stringent specifications for finished fuel blends B5 and B6-20 as well as a standard to control pure biodiesel (B100)[1] product quality prior to blending with conventional diesel fuel.[2]

Alcohol fuels also face hurdles to market penetration. Despite the wide array of alcohols, ethers and other commodity chemicals that can power a flex fuel vehicle, the flex fuel cars sold in the U.S. today are warranted to operate only on ethanol and gasoline. Liquid fuels like methanol and

P-Series (a set of renewable fuel blends that combine ethanol, natural gas liquids and a chemical called methyltetrahydrofuran (MeTHF) which can be made from municipal waste that was approved as alternative fuel by the Department of Energy over ten years ago) are currently excluded from our fuel supply despite their economic competitiveness and the large scale at which they can be produced. If vehicles are not warranted for a particular fuel, there is no incentive for investors to expand production capacity for that fuel. If the proportion of vehicles warranted to use a particular fuel is a very small, only part of the overall vehicle fleet, there is no business reason for fuel station owners to install pumps to serve that fuel.

Scalability is another challenge for alternative fuel advocates. Developers are so in love with their creations that they often forget the limits of feedstock availability or the challenges of collecting and transporting it on a large scale. Indeed, things that look pretty in the lab could be much less so in the real world. We often hear claims that sound like this: "There are X tons of Y in the U.S. This can be turned into Z billion gallons of W, enough to fuel S million cars." It may be true that there are X tons of Y in the U.S. But are they all unused and available as a feedstock for making W fuel? Experience shows that very few feedstocks are not in competition among several end users. The United States has 275 billion tons of coal, which could be converted into 357 billion barrels of equivalent oil products, enough for nearly half a century of current daily consumption. However, coal is used for other purposes, primarily electricity; the conversion of coal to oil is costly, and environmental regulations may make the coal-to-oil path difficult to implement. So while the resource is there, we have to assess its availability realistically. The same is true for other feedstocks. There are 85 million tons of paper and paperboard waste generated in the United States every year. Paper is a form of biomass that can be easily converted into liquid fuel like methanol, but what are the chances that the paper recycling industry, which recycles over 54 percent of the material, would give up on all this paper?[3] The fact that something goes by the name "waste" doesn't mean it's yours, certainly not for free.

This was a painful lesson learned by an ambitious New York-based company called Changing World Technologies (CWT). The company received considerable attention in 2003 for its technology, called thermal depolymerization (TDP), which can take any organic material, break it down into its smallest chemical units and then turn it into oil in a process emulating the earth's natural geothermal activity, whereby organic material is converted into fossil fuel under conditions of extreme heat and pressure over millions of years. All this, according to the company "without combustion, incineration or toxic residue, providing a solution for solid waste management while creating a renewable domestic source of energy." In May 2003, *Discover Magazine* even featured the company in an article titled "Anything into Oil."[4] CWT set its sights on a resource it believed would be both cheap (if not free) and available in mass: offal from meat processing facilities. The company pointed out that if all of this waste were made available for conversion by its technology, the amount of oil generated could almost surpass U.S. oil imports. CWT assumed that concern over mad cow disease would prevent the use of turkey waste and other animal products as cattle feed, and thus this waste would be free. The company teamed up with ConAgra Foods and built a plant next to its turkey processing plant in Carthage, Missouri. As the company later found out, turkey waste may still be used as feed in the United States, and CWT was forced to pay $30 to $40 per ton for it. The lesson: there are no free lunches even if all that is served is turkey offal. Once someone finds a way to assign value to something that is seemingly worthless, it is no longer worthless and therefore no longer free.

The keys to getting to the 48th floor are therefore scalability and profitability. Alternative fuels must be profitable enough to compete in the marketplace, and the focus must be on those alternatives that are sufficiently scalable to take us higher than the mezzanine level.

Abundance of electrons

The fuel that can take us to the highest floor is electricity. In 2008, Americans traveled 2.9 trillion miles on board 230 million vehicles. This alone required 14 million oil barrels per day.[5] How many of those miles were within the range of 20-30 mile trips, the distance that can be covered, if driving a plug-in hybrid, with today's 5-10kWh rechargeable battery? According to the U.S. Census Bureau's latest figures, 137 million Americans commute to work every single working day. Eighty six percent them drive to work, five percent use public transportation, and the rest walk, bike, taxi or work from home.[6] Of those who drive to work 45 percent commute under 20 minutes, 40 percent between 20-45 minutes and 15 percent commute over 45 minutes. Hence, roughly half of car using commuters could do most, if not all, of their daily driving on electrons rather than gasoline, assuming they plugged in their car the night before. If all of those Americans owned PHEVs with a 20-30 mile battery range, 54 million Americans, about the size of the combined populations of New York, Florida and Pennsylvania, would be off oil most days of the year. The liquid fuel tank of the PHEV of course permits a longer journey. But let's face it, how often do we need to go that far? PHEVs offer an estimated gasoline displacement potential of 6.5 million barrels of oil equivalent per day, or approximately 52 percent of the nation's oil imports.

But do we have sufficient electricity generation capacity for so many people to fuel at the socket? Yes, because while our electricity grid is highly burdened during certain parts of the day when usage is at its peak, at night in most of the country there is a great deal of reserve power generation capacity. According to the Department of Energy's Pacific Northwest National Laboratory report, there is sufficient reserve capacity in the electricity grid for up to 84 percent of U.S. cars, pickup trucks, and sport utility vehicles to draw power from the grid during off-peak hours before new power plants of any kind need to be built.[7] But a study conducted by another national lab, Oak Ridge National Laboratory, reveals a more nuanced picture. The authors ran 182 different scenarios for thirteen regions in the United States for 2020 and 2030. They modeled scenarios assuming that vehicles were either plugged in starting at 5:00pm or at 10:00pm and left until fully charged. Three charging rates were examined: 120V/15A (1.4 kW), 120V/20A (2 kW), and 220V/30A (6 kW). It also looked at the generation technology in each region: coal, gas, combustion turbines, steam turbines, or combined cycle plants. The study found that vehicle owners' recharging habits will determine the degree of additional power generating resources needed.[8] Though common wisdom holds that PHEV owners would plug in their car at night with minimal impact on the grid, in reality they may behave differently, plugging their cars in the early evening upon their return from work. In other words, consumers might plug in when convenient, not when utilities prefer. It will therefore be up to the utilities to create incentives and mechanisms to encourage people to do so at the most convenient time. Smart meters can differentiate between plugging and recharging time. In other words, activating the charging mechanism during off peak hours even if the car is plugged in earlier. But even with smart charging we will sooner or later need more power stations. According to the evening charging scenarios, especially by 2030, most regions will need to build additional capacity to meet the added demand from PHEVs. Night recharging, when vehicles are plugged in after 10pm, would require, at lower demand levels, no additional power generation or, in higher-demand projections, just eight additional power plants nationwide. The conclusions of the two studies are encouraging. No doubt there is much to be done to upgrade and secure the American power grid. Such upgrades are needed for a variety of reasons. But at least in terms of power generation capacity, there is sufficient fuel to power a significant portion of the nation's cars and trucks.

Methanol nation

While there is clearly sufficient fueling capacity for those who commute within the battery range of a PHEV, long distance commuters would still need a considerable amount of liquid fuel on a daily basis. But that doesn't mean all of this fuel must be made from petroleum. Assessing America's production capacity of alternative liquid fuels is a complicated task. While we can point to the theoretical resources which the nation and foreign countries can avail for fuel production, it is almost impossible to determine how much of these resources will actually be used for this purpose. Much of it depends on the future price of oil.

We begin with methanol, a politically orphan fuel unlike its cousin ethanol which enjoys at any given time strong Congressional support due to the high number of representatives from agricultural states (in the U.S. Senate there are 60 senators from farm states and only 10 from oil states.) It is not a coincidence that China hedged its bets with methanol. Like the U.S., China has significant coal reserves, and coal is the main feedstock for its rapidly growing methanol industry. But China also has biomass that can just as easily be converted into methanol. Biomass is organic material, such as urban wood wastes, primary mill residues, forest residues, agricultural residues, and dedicated energy crops (e.g. sugar cane and sugar beets,) that can be made into fuel. The raw material is plentiful, and most of the biomass products are currently discarded. Oak Ridge National Laboratory estimated that 323 million tons of cellulose containing raw materials that could be used to create fuel are thrown away each year in United States alone. This includes 36.8 million dry tons of urban wood wastes, 90.5 million dry tons of primary mill residues, 45 million dry tons of forest residues, and 150.7 million dry tons of corn stover and wheat straw.[9] A joint study by the U.S. Department of Energy and the U.S. Department of

Agriculture concluded that the United States can produce more than one billion dry tons of biomass every year and still continue to meet food, feed and export demands. Enough to produce biofuels to supply over one third of the nation's demand for transportation fuels.[10] China's potential is no less promising. Over 80 percent of China's population lives in rural areas. China produces annually more than 800 million tons, mostly in the form of straw, rice husks and wood residue.[11] To understand the potential, consider that according to the Methanol Institute, the production of 10 billion gallons of methanol requires 60 million tons of biomass.

Today, technologies to convert cellulose to ethanol using efficient and cost effective enzymes and other genetically modified biocatalysts are being developed, but these technologies are not yet at large scale commercial production despite the enthusiasm of some politicians and venture capitalists. The attempts to convert cellulose to ethanol by designing bugs that break down wood while secreting alcohol fuel is perhaps the most ambitious genetic engineering project since the Green Revolution in agriculture of the latter part of the 20th century. The process involves a few fundamental phases. Microbes break long chains of cellulose molecules into sugars. The sugar solution is then separated from the residual materials and fermented into alcohol. In nature, these processes are performed by different enzymes and microbes. Cellulosic ethanol visionaries try to create a "bug of all trades," one that can break down cellulose like a bacterium, ferment sugar like yeast and tolerate high concentrations of poisonous alcohol (yes, alcohol is a killer if over consumed). Such a super bug is extremely difficult to create, requiring the manipulation of many genes, but the payback could be fantastic, which explains why so many high profile tech-investors are pouring millions into this burgeoning industry. But biomass feedstocks don't easily give up their starches, and the technologies needed to ferment cellulose in high-pressure chambers that have limited amounts of oxygen are expensive and rudimentary.

While biology is still in its early stages, chemistry is ready for prime time. There are multiple available ways to chemically convert biomass into fuel including combustion, gasification and pyrolysis, each with its own requirement of heat, pressure and time. Perhaps the easiest way to make biomass into liquid fuel is to convert it to methanol. In the first step, a gasifier turns any plant material into synthesis gas consisting of carbon monoxide and hydrogen. This "syngas" is then converted into methanol in a simple chemical process. This technology is mature. After all, methanol is also called "wood alcohol" and has been produced from wood for well over a century. One ton of wood can be converted into 165-185 gallons of methanol. So we can compare apples to apples: that's the same amount of energy as contained in 123-138 gallons of ethanol. If the same ton of biomass were converted into ethanol using an enzymatic process, it would yield only 90 gallons. Coal can also be gasified. Adding an auxiliary unit to integrated gasification combined cycle (IGCC) power plants, which convert coal into synthesis gas, allows utilities to convert part of the gas into alcohol fuels and sell both electricity and liquid fuels. One of the Department of Energy's clean coal demonstration program's most successful efforts is a commercial scale facility in Kingsport, Tennessee that generates methanol from coal at roughly 50 cents a gallon. Methanol contains about half the energy of gasoline per gallon so that's equivalent to about one dollar for a quantity of methanol that will take you as far as one gallon of gasoline. Producing one million gallons of methanol requires about 5,000 short tons of coal. So 4 percent of current U.S. annual coal production, which in 2007 was 1,146 million short tons, would yield 10 billion gallons of methanol, which is about the same amount of fuel the corn ethanol industry contributes today to America's fuel supply.[12]

There are many other ways to produce methanol. In Germany, Schwarze Pumpe produces 100,000 tons of methanol from sewage sludge and industrial wastes each year. In Sweden, methanol is made from black liquor, a sludge byproduct of paper pulping. Natural gas can also be a feedstock. About a third of the world's emissions of methane, a greenhouse gas 23 times more potent than CO_2, occur in coal mines and natural gas wells, where billions of cubic feet of natural gas are currently being flared by oil companies. According to the World Bank sponsored Global Gas Flaring Reduction Partnership, the total amount of five trillion cubic feet of natural gas are being flared annually, equivalent to 27 percent of total U.S. natural gas consumption. Gas flaring from Nigerian wells and refineries alone emits more greenhouse gases than any other single source in Africa south of the Sahara. Instead, much of this flared gas could be turned into valuable methanol and used as a petroleum replacement on board flexible fuel vehicles. It takes about 100 cubic feet of natural gas to produce one gallon of methanol. In Equatorial Guinea, for example, 100 million cubic feet per day of natural gas that had been previously flared are being turned into some 300 million gallons of methanol per year. This can, and should, happen elsewhere. Using about 10 percent of the world's flared natural gas would produce methanol to fuel five million cars. But there will not be a market incentive for any of this unless we have methanol enabled cars on the road. Unfortunately, the flex fuel cars sold in the United States are warranted to run only on ethanol. Barring an Open Fuel Standard which enables *all* alcohols to compete at the pump, many billions of gallons of alcohol fuel that could displace petroleum will be excluded from the marketplace, and our alternative liquid fuel market will forever remain at the mercy of the corn ethanol sector, subject to the whims of Mother Nature and agribusiness lobbyists.

Ethanol: the fuel pundits love to hate

Which brings us to ethanol. As should already be clear, alcohol doesn't just mean ethanol and ethanol doesn't just mean corn, but in the United States, corn ethanol has been for decades at the center of a huge political controversy. America is essentially divided into two camps: corn ethanol haters and corn ethanol fans. The corn ethanol industry is well organized, well represented politically, and by virtue of Iowa's influence on presidential politics enjoys significant political power. As Senator John McCain learned in his 2000 presidential campaign and Mayor Rudy Giuliani in 2008, the road to the White House passes through Des Moines, and singing ethanol's praise is part of any candidate's job description, whether Republican or Democrat. The agricultural subsidies the U.S. ethanol industry and the blenders who blend it with our gasoline secured themselves have caused many Americans to identify corn ethanol with pork and corruption rather than with energy security. "Scam," "fraud," and "boondoggle" are some of the adjectives used to describe corn ethanol. "In a capital city that is full of shameless political scams, ethanol is perhaps the most egregious," wrote Kevin Hassett of the conservative American Enterprise Institute.[13] "Ethanol is one of the shameless energy rackets going, in a field with no shortage of competitors," concluded the *Wall Street Journal* editorial board in a March 2009 editorial titled "Everyone Hates Ethanol."[14] The opposition to corn ethanol no longer contains itself to the grounds of undue government intervention in the free market and abuse of taxpayers' dollars. Ethanol has been blamed for almost all of the world's problems from starvation to land depletion to global warming. In a spectacularly erroneous *Time Magazine* cover story, Michael Grunwald wrote about the "clean energy scam" blaming ethanol and other biofuels for "dramatically accelerating global warming, imperiling the planet in the name of saving it."[15] Others see ethanol as an enemy of the poor. Lester Brown of the Earth Policy Institute alleged that biofuels pit the 800 million people who own automobiles against the world's two billion poorest people.[16] Statements to that effect coming from American reporters and pundits are music to the ears of some of America's critics abroad. A U.N. special rapporteur on the right to food, Jean Ziegler, called biofuels a "crime against humanity,"[17] Cuban leader Fidel Castro called the Bush Administration's ethanol project "genocidal," and his protégé, Venezuela's Hugo Chavez, referred to it as "true madness."[18] Unsurprisingly, the Islamists have also put ethanol in their sights. Mohammed al-Najimi, a prominent scholar at the Saudi Islamic Jurisprudence Academy, warned students traveling outside Saudi Arabia not to drive any vehicles powered with ethanol "because the prophet has cursed not only who drinks it but also those who use it for other purposes."[19] Astonishingly, ethanol was even blamed for driving up the price of oil. OPEC's president alleged that 40 percent of the rise in oil prices can be attributed to the "intrusion" of ethanol.[20] In the summer of 2008, as oil price hit peak after peak, ethanol, the only alternative fuel that was somewhat able to keep the lid on gasoline prices, was facing the most ferocious attack ever.

Let's take a deep breath, deconstruct the debate, separate the fallacies from the truth and try to squeeze out some of the venom. But first things first. We are opposed to agricultural subsidies of any kind. For corn, for sugar, for soybeans, for anything. U.S. agricultural policy is a sham. We have sugar quotas that keep U.S. sugar prices artificially high above world prices – kept in place by a sugar lobby that gives equal opportunity campaign contributions to both Democrats and Republicans – and at the same time for years, federal agricultural policy kept corn prices artificially low through a price support system that disconnected corn price from actual demand. So while around the world sugar is used to sweeten baked goods, cereals, and so forth, in the United States, food manufacturers became accustomed to using cheap – taxpayer subsidized – corn syrup. No wonder obesity and juvenile diabetes are plaguing our society. We pay farmers not to farm, and we have ridiculous import tariffs and trade barriers on imported products. These are all wrong. But let's separate all that from the merits of ethanol as a viable transportation fuel. Ethanol is not a perfect fuel. Far from it. It packs less energy than gasoline, it depends on the volatile agricultural sector and, like other alcohols, it is more corrosive than petroleum fuels. If gasoline were free of problems, there would be no need for ethanol, and probably not for this book. But it isn't, and our oil dependence is costing us more and more every year. Which is why our judgment on ethanol should be based on one consideration: are the security and economic costs of ethanol lower or higher than those of oil?

Billions of dollars of various subsidies and tax credits are allocated to the U.S. ethanol industry annually despite that fact that when oil prices surpass $45 a barrel, corn ethanol is competitive with gasoline on a per mile basis without subsidies, as long as corn is under $3 a bushel. But the wanton allocation of taxpayer dollars to favored energy sectors is certainly not confined to ethanol. How much taxpayer money is given to oil? Or nuclear? One could spend a lifetime reviewing studies that looked at the issue, but here is a sample. According to Earth Track, an organization dedicated to research on energy subsidies, in 2006, ethanol subsidies stood at $6 billion, making 7.6 percent of the total federal fiscal subsidies to energy, not a trivial amount considering the industry's relatively small size. In the same year, the nuclear industry was on the receiving end of $9 billion and the oil and gas industry received $39 billion, about 52.4 percent of total energy subsidies.[21] Other studies show much higher figures for oil and gas subsidies. One, co-authored by Republican C. Boyden Gray, who served as White House counsel to President George H. W. Bush, factored in air quality costs alleging that the oil industry is subsidized to the tune of $250 *billion* a year![22] Studies that assessed the cost of our military capabilities necessary to defend the flow of oil from the Persian Gulf point to a "hidden subsidy" of $50-60 billion annually in a non-war year − we pay for this through our income tax, not at the pump. We aren't subsidizing the defense of semiconductor shipments from Taiwan. Choose any number you want and the verdict is clear: energy subsidies are not unique to ethanol; they are unfortunately prevalent throughout our energy system (not to mention pervasive throughout our economy.) But when assessing ethanol's economic cost to our society one must keep one thing in mind: On a gasoline equivalent basis, ethanol's contribution to the U.S. fuel market is only second to Canada. Equivalent to about 750,000 barrels of gasoline per day, it surpasses that of Saudi Arabia, Mexico and Venezuela. This means that if not for ethanol, our oil demand would have been much higher and with it our trade deficit. Ethanol is a domestic fuel and the money spent on the industry is handed to hard working, tax paying Americans who plough most of it back into the U.S. economy. With oil imports, on the other hand, we export our wealth to the economies of some of the worst regimes in the world. Furthermore, according to Merrill Lynch, by adding supply to the fuel market and thus taking off some of the pressure at the height of oil prices in the summer of 2008, ethanol was responsible for keeping the price of oil 15 percent lower than where it would have otherwise been.[23] This means that annoying as the $6 billion ethanol subsidies might be for many (including us), by keeping the lid on oil prices the use of ethanol saved the U.S. economy in 2008 more than $60 billion – ten times the subsidy – that otherwise would have ended up in the coffers of foreign oil exporting countries. Not a bad deal.

One of the mantras of ethanol opponents is that it takes more energy to make ethanol from corn than you get when you burn it in your car. This energy balance argument is both irrelevant and intellectually dishonest. Irrelevant because it is a basic law of nature that when converting raw energy (like crude oil, coal, corn or sugar cane) into a usable form (like gasoline, electricity, or ethanol) you always have to put more energy in than the energy you get out: the energy inherent in the matter, plus the energy of conversion. If we were to make our choices on the basis of energy balance considerations, the first source of energy to relinquish would be our food. As food expert Michael Pollan noted, it takes 10 calories of fossil-fuel energy to produce a single calorie of modern supermarket food.[24] We still eat three meals per day plus snacks. Intellectually dishonest, because net energy proponents never mention the amount of energy that goes into making a gallon of gasoline. Think how much energy is needed to drill oil in Saudi Arabia and ship it all the way to the United States. Pumping seawater into the wells to increase reservoir pressure, pumping the oil out of the well, transporting the crude to processing facilities where sulfur and other impurities are removed, loading the oil on a tanker, powering the tanker in a one-month voyage across two oceans, refining the oil into gasoline and shipping the gasoline to gas stations throughout the United States all require energy. A lot of it. The amount of fossil fuel in mega joules needed to make one mega joule of gasoline is 1.19 versus 0.77 for corn ethanol and 0.10 for cellulosic ethanol.[25] The point is: gasoline's energy requirement is greater than ethanol's (unless we count the energy of the sun absorbed by corn plants, which would be a pretty odd thing to do given that the sun doesn't charge) and yet this hasn't prevented any of us from using this "net energy loser."[26]

Equally hollow is the argument that ethanol's mileage per gallon is lower relative to gasoline. The issue is not *miles per gallon*. If one is comparing the fuels on a cost basis, the relevant comparison is *cost per mile*. Gallons of ethanol, methanol, gasoline, and other fuels all contain different amounts of energy and thus will carry a given vehicle a different number of miles. If oil is above $45 a barrel, corn ethanol is competitive with oil on a per mile basis. If one is looking at it from an energy independence perspective, the relevant metric is *miles per gallon of petroleum fuel*. Precious little petroleum goes into the making of even corn ethanol. The primary energy inputs into the production of corn ethanol are the coal that powers the plants, the natural gas from which fertilizers are made and the solar energy absorbed by corn plants.

Another arrow in the quiver of the anti-ethanolists is the allegation that ethanol causes more greenhouse gases than the gasoline it replaces if indirect carbon dioxide emissions are taken into account. Indirect analysis thus counts the carbon dioxide released when farmers in the developing world supposedly plow and burn forest or grassland to grow energy crops used to make biofuels or to grow food to replace previously imported crops whose supply was reduced by the demand for ethanol. Particularly influential was the article by Timothy Searchinger, formerly staff attorney at an environmental organization, in *Science Magazine* in which he argued that U.S. ethanol production, by reducing dumping of subsidized U.S. food on the global market, is driving agriculture to expand in the developing world, according to Searchinger in significant part by burning down forests and plowing up grasslands, a practice that ends up releasing twice as much carbon dioxide to the air as the original amount of displaced gasoline would.[27] Putting aside the criticism of Searchinger's article by scientists who pointed out that it is exceedingly difficult to actually measure how much if any deforestation abroad is specifically due to ethanol consumption in the U.S., as Robert Zubrin has eloquently written, his argument implies that economic growth in the developing world is not desirable because that growth comes with increased emissions, and that indeed economic growth in the U.S. is a bad thing too because "anything that allows Americans to buy increased quantities of groceries can be said to cause deforestation and thus global warming."[28] This is a profoundly anti-human view. Nevertheless, the *Science Magazine* article poked a big hole in environmentalists' traditional support for ethanol as a green and renewable fuel. Inspired by such arguments, the California Air Resources

Board (CARB), the same agency that in 2003 gained notoriety for its role in "killing" the electric car, proposed a Low Carbon Fuel Standard that would penalize ethanol importers and producers for indirect greenhouse gas emissions supposedly resulting from deforestation and plowing up grasslands. According to this proposal, an importer of ethanol from, say, the Dominican Republic, would be liable for any greenhouse gas emissions caused by the alternative land use.

Deforestation is a global problem, and in some parts of the world it is indeed tied to biofuels. Indonesia, which cleared its forests to create palm plantations for biodiesel, is one such case. But what Searchinger and his colleagues failed to notice is that in most cases, deforestation has nothing to do with ethanol. It's primarily caused by timber production and cattle ranching. The cutting down of the Amazon rainforest has nothing to do with the ethanol industry. In fact, the Amazon's climate is not even suitable for growing sugar cane. Brazil is the world's biggest producer and exporter of sugar cane-based ethanol. Sugar cane occupies only 6 million hectares of the 62 million hectares of Brazil's cultivated farmland. Roughly half the cane is used to produce ethanol and the rest is for sugar. There are another 90 million hectares of degraded pasture that can be used for farming without even going near the Amazon. The economic signals driving biofuels or agricultural land-use are divorced from the timber-driven economic signals that cause the changes in land-use. If we are to penalize anybody for deforestation let's first take a hard look at the origin of our furniture, newspapers, books, and the 24 rolls of extra-soft bathroom tissue most of us flush down each year. Searchinger and his ilk are equally insincere for failing to apply similar accounting for indirect greenhouse gas emissions resulting from the use of gasoline such as the military operations related to defending our access to oil (cf. the First Gulf War). After all, the jets, tanks, ships and Humvees patrolling the Persian Gulf or the Special Forces protecting the oil pipelines in Colombia don't run on vegetable oil, and the electricity powering military bases dedicated to protecting our access to oil is not made on wind farms. Ignoring those while speculating about the role of deforestation is selective bias, not science. Michael Wang from the Center for Transportation Research at the Argonne National Laboratory who applied a vastly different and more realistic methodology found that on a per gallon basis, when the full "life cycle" of the fuel, from growing it (or recovering it from the ground in the case of oil) to producing the fuel and burning it, is taken into account, corn ethanol reduces greenhouse gas emissions by

18 percent to 29 percent compared with gasoline, sugarcane ethanol reduces emissions by 56 percent and cellulosic ethanol has an even greater benefit with more than 80 percent reduction.[29] Another report commissioned by the International Energy Agency (IEA) examined greenhouse gas (GHG) reductions from grain ethanol since 1995 and concluded that thanks to improvements in both feedstock production and ethanol production, GHG reductions have grown from approximately 26 percent in 1995 to over 39 percent today, while projected GHG reductions from ethanol will reach nearly 55 percent in 2015.[30]

The food vs. fuel myth

Searchinger was one voice in the choir of analyses and reports that inundated the media when oil prices were at their peak, with claims that the U.S. corn ethanol effort is affecting food production around the world and hence starving poor people. It seemed so obvious, the headlines blared, that with so much corn being turned into fuel, food shortages must inevitably result, and so if there is a food shortage anywhere, it must be because of America, its profligate motorists and its corn ethanol program. Contrary explanations for the rise in food prices like the role of speculators, the rise in oil prices and the fact that hundreds of millions of people in China and India are rising out of poverty and moving from mere subsistence to a more calorie rich diet and hence demanding more meat which put enormous pressure on the grain market (it takes eighteen calories of grain to produce one calorie of meat) were ignored by the pundits who eagerly jumped on the anti-ethanol bandwagon. How could that drastic increase in the price of food commodities from fish to rice possibly be attributed to ethanol? Nobody was growing corn in rice paddies or making biofuels out of fish. The economic crisis came, and proved that all this was a farce. Food commodity prices track oil prices regardless of how much corn is used for ethanol production. When oil was up, it affected the cost of essential components of our food supply chain like fertilizers and transportation. When prices came down, food prices quickly followed. Between July and November 2008, oil fell nearly 50 percent. In the same period of time, corn prices fell by the exact same figure. Was this because we used less ethanol? No. To the contrary, the U.S. Agriculture Department reported a 23 percent increase in farm-grown exports in 2008 over 2007, and ethanol production capacity increased by 60 percent to 11 billion gallons, saving the global oil market almost one million barrels of oil per day. America is clearly doing its share in both feeding and fueling the world. Grain prices were driven up by the same speculative forces that jacked up oil prices. When the bubble created by commodity speculators popped, prices declined. This is not to say that ethanol had no impact at all on food prices. It did. But this impact was marginal in comparison to the other factors in play. A 2009 study by the Congressional Budget Office (CBO) found that increased corn ethanol production accounted for only 10 to 15 percent of total increased food costs in the period between April 2007 to April 2008. Higher corn prices accounted for 0.5 to 0.8 percentage points of the 5.1 percent increase in food costs as measured by the consumer price index. "Over the same period, certain other factors – for example, higher energy costs – had a greater effect on food prices than did use of ethanol as a motor fuel," the CBO determined.[31] The reason the impact was so minimal was because the cost of food ingredients in food products we buy at the supermarket represents only one fifth of the price at checkout. When it comes to corn, the cost of the ingredient is much lower. At $5 per bushel, a $3.00 box of cornflakes, which contains 15 oz of corn, contains corn that cost 8 cents when bought from the farmer. So farm commodity prices have almost no effect on the retail consumers. But the effect of oil price hikes can be huge – and not just on food, but on all consumer goods requiring fuel for processing or transport. Interestingly, despite the drop in food commodity prices, retail processed food prices have not dropped. We don't hear anyone accusing "Big Food" of price gouging though. "Big Food" through hired lobbyists and PR firms has done quite a good job of deflecting blame to the ethanol industry. These lessons are worth remembering as sooner or later the economy will pick up steam and oil prices will most likely rebound, perhaps even to a much higher level. At this point the anti-ethanol coalition will no doubt regroup to resume its effort. When this happens, let's make sure to remind them what the substantive causes of high food prices are. In the interim, rather than shut down the biofuel programs, we need to radically augment them, to the point where we can take down the oil cartel.

Despite of all of the above arguments in defense of corn ethanol, one should not be misled to believe that this fuel alone can bring us to the Promised Land. It can't even under the most optimistic scenarios. Ethanol yield per bushel of corn in the United States has increased significantly since 1980 and is likely to continue to do so in the years to come. This is partly due to the increase in corn yield per acre – in 2007, corn growers produced on average 151 bushels per acre; according to the ethanol industry, this number is projected to grow to 211 by 2018 – and partly due to improvement in refining techniques. But no matter how efficient the industry becomes, corn will likely forever be a relatively low yield crop. Currently corn yields 300-400 gallons per acre per year. Compare this to 800 gallons with sugarcane and 1,000-1,500 gallons in the case of cellulose. There are even bigger bonanzas. Technologies to convert seaweed to ethanol are projected to yield as many as 2,500 gallons per acre per year. The reason for this astounding yield is that seaweed grows at a phenomenal speed essentially doubling its size every few hours.

Our biodiesel industry also suffers from reliance on low-yield crops. Today, most of our biodiesel is made from soy at a yield of 48 gallons per acre per year. In Asia, where biodiesel is made from palm oil the yield is over 600 gallons per acre. But new processes to convert oil-rich micro-algae to biodiesel can yield as much as 5,000 gallons. This early stage technology is particularly interesting as it involves injection of CO_2 into the algae ponds and is highly efficient (more about this in the next chapter). According to one algae company, "since the whole organism converts sunlight into oil, algae can produce more oil in an area the size of a two-car garage than an entire football field of soybeans."[32]

The scandalous ethanol tariff

The point to remember is, to reiterate, that ethanol is not the only alcohol and corn is not the only source of ethanol. The greater productivity of cellulosic sources should eventually allow them to produce enough alcohol to displace nearly a third (90 billion gallons) of all gasoline use by 2030 according to a 2009 study released by Sandia National Laboratories and 150 billion gallons by 2050, according to a report by the Natural Resources Defense Council (NRDC).[33] Most of these gallons will come from cellulosic feedstocks such as switchgrass, corn stover, wheat straw and woody crops. That's the equivalent of more than two-thirds of the current gasoline consumption in the U.S. All this assuming that gasoline prices return to their pre-economic crisis upward trajectory. But until the transportation fuel market is opened up to competition, the U.S. ethanol industry will be growing at the pace dictated by a Soviet-style central planning Renewable Fuel Standard (RFS) passed by Congress in 2007. The RFS set a detailed year by year mandate on the level of renewable fuel in the market. It is certainly folly to imagine that any number of bureaucrats, regardless of their training, would have the ability to determine the optimal alternative fuel level in any given market at any given year. That is why an Open Fuel Standard is necessary. An Open Fuel Standard, by simply ensuring vehicles have a minor $100 tweak enabling them to be powered by a any combination of liquid fuels made from a variety of feedstock, something that automakers have agreed in principle to do and now simply need to be buttressed by law, would, by putting the fuel choice in the hands of the fuel buyer, enable the market rather than government bureaucrats to determine the optimal level of alternative fuel in the market as a function of the myriad of factors that impact consumer decisions, chief among them generally being comparative price. This is clearly illustrated in Brazil: when oil prices go up, drivers put more alcohol in their tank, and vice versa. (It is interesting to note, that some of the members of Congress and administration officials most supportive of the RFS declined to support the OFS calling it a mandate. Our government officials have not been lately blessed with an inclination to intellectual consistency.)

The Open Fuel Standard will also enable us to pit sugar ethanol against gasoline, as is done in Brazil. Unfortunately, the United States is not able to ramp up sugar production to the level allowing it to implement the Brazilian model. Sugar needs a long, frost-free growing season, and expansion of sugar growing beyond Florida, the Gulf Coast and Hawaii is limited. The good news is that many other countries in Africa, Latin America and South Asia can, and we can import the fuel from them. Brazil, the Saudi Arabia of sugar, already exports half a billion gallons of ethanol a year. "We don't want to sell liters of ethanol, we want to sell rivers," Brazil's Agriculture Minister Roberto Rodrigues said in 2004. Brazil can no doubt do that. It currently uses less than one percent of its total territory, and its sugar cane production is concentrated mainly in two geographic regions – the Northeast and the Center South, both a considerable distance from the Amazon rainforests. Expanding U.S. fuel choice to include biofuels imported from foreign countries also has geopolitical benefits. Sugar is now grown in 100 countries, many of which are poor and on the receiving end of U.S. development aid. Encouraging these countries to increase their output and become fuel suppliers could have far-reaching implications for their economic development. By creating economic interdependence with its neighbors in the Western Hemisphere, the United States will guarantee that those poor countries do not fall on the side of China which has already set its sights on Western Hemispheric energy supplies or, worse, Hugo Chavez's so-called "anti-imperialist" block. Yet, in the United States, the ethanol and sugar industries oppose imports of sugar ethanol and got their champions in Congress to impose a stiff tariff of 54 cent per gallon on imported ethanol to protect the local industry. On a gasoline equivalent basis, this tariff is equivalent to a tax of roughly $20 a barrel on imported oil. Indeed, we don't tax oil imports from unfriendly countries like Saudi Arabia and Venezuela, but we do tax alternative fuels coming from some of the friendliest sources. Despite protestations of ethanol champions that lifting the tariff would be "a kick in the face to rural America where the ethanol comes from," blocking ethanol imports to the United States to protect corn is a scandalous practice that puts into question the corn ethanol industry's sincerity when invoking U.S. national security.[34] It is tantamount to blocking gasoline imports to protect domestic gasoline producers. Such policy makes ethanol protectionists in Congress the biggest obstacle for full-scale deployment of ethanol in the United States. Any reasonable policymaker should see beyond the local political expediencies and recognize that we'd be far better off importing ethanol from our friends than oil from our enemies. But with the introduction of an Open Fuel Standard, the domestic industry's opposition is likely to wane. With tens of millions of flex fuel vehicles on the road, the fuel market would simply be too big for the domestic corn ethanol industry to supply and therefore there would be no grounds for its fear of foreign competition. We've never heard the domestic oil industry complaining about foreign competition. Simply put, in a big enough market there is room for everybody. And everybody means everybody, not just oil or corn. Be it coal, algae, seaweed or garbage, the United States is blessed with plenty of feedstocks from which alternative fuels can be made. Not all of the alternatives are competitive with oil using today's technology, especially when prices are low, but many are, and they could demonstrate their competitiveness were they allowed to enter the marketplace. Others will become competitive as oil prices climb again as they surely will. But as long as our cars are able to run only on gasoline these solutions will not amount to much, and the monopoly of oil in the transportation sector and with it the excessive power of the oil cartel will continue to prevail.

⌘ ⌘ ⌘

6
MELTING ICE MEETS MELTING WEST: POWERING OUR CARS WITH CO_2

In the short history of the 21st century, no issue has risen from near obscurity to the center of our public discourse as quickly as global warming. What started as chatter among some concerned scientists and diehard environmentalists within just a few years was being called "the challenge of our generation," consuming the bandwidth equivalent of war in terms of attention of governments and world leaders. Greenism is one of history's fastest growing mass movements, and for the first time since the rise of Communism, millions are coming together behind an idea that through a fundamental change of our economic order – in the case of the green movement through a sharp cut in greenhouse gas emissions – we could solve a slew of global problems from social injustice to poverty and infectious diseases.

Like other mass movements in history, the green movement fosters a narrative about impending catastrophe should we maintain a business-as-usual approach to energy and continue to emit increasing amounts of heat trapping gases. Doomsday scenarios of rising sea levels, violent storms, famine and starvation, infectious diseases and extinction of species have created massive anxiety among millions. Adults increasingly view global warming as the biggest threat to the future of humanity, while today's children are more worried about inundation by water of their neighborhood than about auto accidents, cancer or terrorism.[1] Much of the hysteria has been fanned through a mobilization of the media. As early as 2005, *USA Today* announced: "The Debate's Over: Globe *is* Warming."[2] In April 2006, a *Time Magazine* cover screamed "Be Worried, Be Very Worried."[3] Two years later, the same magazine featured a cover story titled "How to Win the War on Global Warming," showing an altered image of Joe Rosenthal's inspiring World War II photo of U.S. Marines raising the flag on Iwo Jima, in which the flag was replaced with a tree. *Time*'s message was clear: we are at war, and we must win if we are to survive.

But there is also another war going on. Green may be the color of environmentalism, but it is also the color of the flag of the Islamists who wish us dead. Green is also the color of our currency, the dollar, which is facing grave danger as a result of our economic decline, the mounting national debt and our growing trade deficit, largely caused by our thirst for oil imports. So while there is no lack of evidence that the icecaps are melting, the West too is melting, culturally as well as economically, and at no slower a pace. This is partly due to the failure of Western governments to stop the spread of radical Islam into the heart of their societies which is largely enabled by the transfer of wealth from oil consumers to oil exporting countries. A lot has been written about the role of petrodollars in funding radical Islam and shifting the balance of power within Islam from moderates to radicals. Lawrence Wright's widely quoted statement that Saudi Arabia's Wahhabis who barely account for one percent of the world's 1.5 billion Muslims fund 90 percent of the expenses of the entire faith is particularly illuminating.[4] It means that the most radical streak within Islam is funding the construction of nine out of ten mosques, the salaries of nine out of ten imams and the publication of nine out of ten books on Islamic affairs. "Fifty years from now," Thomas Friedman wrote, "we may conclude that the most important geopolitical trend [of our time was the] shift of the center of gravity of Islam away from Cairo-Istanbul-Casablanca-Damascus/urban Mediterranean center of gravity in the nineteenth and twentieth centuries, which tended to be softer-edged, more open to the world and other faiths, and toward a Salafi/Saudi desert-centered Islam which was much more puritanical, restrictive toward women and hostile to other faiths."[5] Such an oil fueled shift from the moderate to the radical form of the world's fastest growing religion has profound implications for the future of the West, as radical Islam sends its tentacles deeper into our civilization.

Let's face it. If there is any hope for significant greenhouse gas reduction, it comes from the West. Europe and the United States for all their tactical differences on how to achieve this goal are the main drivers of the so called green economy. Without the West, Al Gore's battle is lost. The salvation certainly won't come from developing countries where planetary concerns take a lower priority than basic human needs – access to electricity, food, and shelter. Furthermore, any success in curbing greenhouse gas emissions would be contingent on Western prosperity. An economically depleted West doesn't bode well for the climate movement, and such prosperity would be difficult to achieve as long as Western economies bleed due to the need to import increasingly expensive oil. So if we are to make real progress toward energy independence and reduced greenhouse gas emissions, it will take a unified front of champions of the environment, national security and a strong economy all working in sync to ensure that oil no longer holds us over a barrel.

What's between tree huggers and cheap hawks?

This was the rationale behind the founding in September 2004 of the Set America Free Coalition (www.SetAmericaFree.org). This unique strange bedfellows alliance of "tree huggers, do gooders, sod busters, cheap hawks and evangelicals," as founding member James Woolsey likes to call it, has taken upon itself to find common ground among people who may not agree with each other on almost anything. From uber-conservatives like evangelical leader Gary Bauer and security hawk Frank Gaffney to Obama advisors and cabinet members like Senator Tom Daschle and Secretary of Interior Ken Salazar, from environmental and labor groups like the Natural Resources Defense Council and the Apollo Alliance, to veterans, prisoners of war and 9-11 families, members of the coalition are united in their belief that breaking oil's transportation fuel monopoly is a win-win-win solution: it is good for national security, good for the environment and good for the economy. Yet, despite the collaborative spirit, sometimes energy security concerns breed policies that environmentalists consider unacceptable and vice versa.

A few examples: During the apartheid years, South Africa faced economic sanctions, which threatened its oil imports. The country addressed its energy challenge by building coal-liquefaction facilities. Today, coal-rich countries like China and the U.S., eager to cut petroleum dependence, are increasingly interested in similar coal-to-liquids technology, which is profitable as long as crude oil remains above $60 a barrel. But, for environmentalists, using coal to displace oil is a nightmare scenario, as coal-derived gasoline emits twice as much CO$_2$ per mile on a lifecycle basis than petroleum-based fuel. Coal is not the only source of energy that improves energy security while increasing CO$_2$ emissions. Canadian tar-sands deposits are estimated at 1.7 trillion barrels of crude oil, but the environmental impact of extracting them far exceeds that of conventional oil. Indonesia presents another example. After recently turning into a net oil importer, the country began burning rainforests to create palm-oil plantations for biodiesel. This caused such an increase in CO$_2$ emissions that the country turned into the world's third biggest emitter after China and the United States. There are also examples of policies that address environmental concerns but compromise national security. The prime exhibit here is Germany. Chancellor Angela Merkel named confronting climate change as her country's top priority and has already succeeded in getting the 27 European Union governments to agree to collectively cut greenhouse-gas emissions by 20 percent by 2020. Spearheading the effort, the German government announced that it will seek to totally phase out the country's entire coal-mining industrial sector by 2018. Germany also intends to phase out its nuclear-power industry by 2020 (this despite the fact that nuclear power plants do not emit CO_2). These are astonishing decisions considering that 52 percent of Germany's electricity is generated from coal and 29 percent from nuclear energy. Germany is taking a huge gamble – verging on recklessness – by attempting to rid itself of so much of its base load power-generation capacity and replace it with Russian natural gas and a slew of renewable-energy technologies like solar and wind which in the absence of a huge leaps forward in grid scale energy storage technologies are intermittent (the wind doesn't blow 24 hours a day seven days a week nor does the sun shine all the time) and require government subsidies to compete. In the name of earth stewardship, the German government has decided to put its economy at the mercy of Vladimir Putin, who has thus far shown no compunction in using energy as a geopolitical weapon.

The biggest challenge in squaring security and environmental considerations involves India. This huge country's growing demand for electricity puts it on the horns of dilemma: as home to 10 percent of the world's coal reserves, it could provide for most of its own power needs. Coal power for one billion Indians means a lot of CO_2. Yet, security-minded people are even more concerned about India shifting to the cleaner alternative to coal: natural gas. Should India decide to power its turbines with natural gas, it is likely to become increasingly dependent on nearby Iran, the world's second largest natural gas reserve, which is eager to build a natural gas pipeline through Pakistan to India. Pressuring India to reduce its emissions may slow down the melting of the ice-caps, but such a policy will expedite the melting of the West by sending India right into the welcoming arms of Iran's mullahs, undermining Western efforts to isolate Iran economically.

All of these cases show how difficult the tradeoffs between environmental concerns and energy security can be. The good news is that there are policies and technologies that could successfully address both, among them increased efficiency, conservation and renewable energy technology.

Fuel choice is perhaps the best policy that could bring greens and hawks together. But for this we need to think a little out of the box.

Bury it for later

The box is currently dominated by an idea called carbon sequestration. The goal of this lavishly funded approach – really a hydrogenesque boondoggle – is to capture CO$_2$ collected in power stations and inject it into deep underground rock formations and storage sites with the hope that it will stay there forever. The concept is essentially to treat CO$_2$ as a hazardous waste and bury it underground or underwater. Just like in the case of hydrogen, to the unscientific ear this sounds like a neat idea, but in reality, other than in places where injecting CO$_2$ would have economic utility, for example in aging oil wells to enhance oil recovery, carbon sequestration is an idea devoid of scientific and economic merit. According to the Energy Information Administration (EIA), in 2007 the United States emitted nearly 6 billion metric tons of CO$_2$. This means that in volumetric terms, in 2007 the United States emitted the equivalent of four times the total mass of coal and oil extracted from its ground. Approximately 30 million metric tons of the emitted CO$_2$ are injected into declining oil fields in the U.S. annually, which amounts to about half a percent of overall U.S. CO$_2$ emissions. Putting aside for a moment the question of efficacy, injecting into the ground or water for the sole purpose of disposal a volume of gas of the magnitude of U.S. CO$_2$ emissions would impose a huge cost on consumers. The effort involved in constructing a network of pipelines to carry the gas to its sequestration sites and the energy required to transport and inject the CO$_2$ would be monumental. In the coming decade, more than 800 new coal-fired plants will be built in China, India and the U.S. alone, pumping up to five times as much CO$_2$ into the atmosphere as the Kyoto Protocol aimed to reduce. It is difficult to see how most of these power stations, many of which have already been sited, could be co-located with sequestration sites. It is unlikely geology would cooperate, so there would be no avoiding building pipelines to send the gas far away for disposal. The added cost to the consumer would be eye popping. Aside from the cost issue, there are many technical and safety problems that make sequestration a questionable approach. CO$_2$ is a leaky gas that tends to migrate to the surface. Sequestration proponents promise that leakage would be minimal. The Intergovernmental Panel on Climate Change (IPCC) estimates that well selected stores are likely to retain over 99 percent of the injected CO_2 over 1,000 years, but such assertions are no more reassuring than those coming from mobile phone companies telling us that there is no long term link between cell phone use and brain cancer. Even an MIT report on the future of coal which overall was bullish about the prospect of sequestration admitted that it is not yet possible to provide quantitative estimates of the possibility of leakage from storage sites and that high CO_2 concentrations could cause adverse health, safety, and environmental consequences.[6] A leakage of as little as two percent per year means that within 50 years most of the stored CO_2 will be back in the atmosphere, making sequestration an expensive feel-good solution but one that will have minimal impact on long term dialing back of carbon dioxide emissions. Sequestration can also be dangerous. If an underground CO_2 bubble slips to the surface as a result of a weak earthquake the outcome could be devastating. In 1986, a gigantic natural CO_2 bubble which emerged from Lake Nyos in Cameroon when gas saturation levels in the lake surpassed a critical level caused the death of 1,700 people and 3,500 livestock. One can only imagine what could be the impact of such release in the seismically sensitive Northern California.

The new garbage

The challenges associated with sequestration beg for an alternative approach. To understand what our approach toward CO_2 should be, it's worth taking a look at our evolving approach toward garbage. In the past, all of our garbage was collected and shipped on board trucks and barges to distant landfills where it was dumped and incinerated, releasing toxic particles into the air in the process. Over time, the landfills moved further and further away from cities, and the cost of disposal grew by leaps and bounds, making garbage recycling an economically viable option. Today, we still send millions of tons of waste to landfills, but many of us no longer send our newspapers, cardboard boxes, aluminum cans, glass jars and plastic bottles to the landfill. We learned to recycle them. The beauty of recycling is that it treats what was formerly considered waste as an asset rather than a liability. Your plastic bottle – which happens to be made from oil or natural gas – is no longer a piece of garbage that needs to be burned; it is a raw material that, when the economics make sense, can be used for a variety of end products, including sleeping bags, fleece jackets, packaging material, fences, park benches, signposts, drainage pipes, electrical fittings and floor covering. In other words, the empty bottle now has inherent value. Similar thinking should guide us in dealing with carbon dioxide. Instead of focusing on ways to dump the stuff at high cost, we should think of CO$_2$ as a raw material that can be made into usable products. Indeed, CO$_2$ can be incorporated into polymers and converted into a variety of chemicals. But there is a limit to how much of most industrial chemicals we can use. The largest potential market for CO$_2$ derived products is the transportation fuel market, and that's where the focus should be.

There are a number of approaches here. Nobel Laureate in Chemistry Professor George Olah has been a proponent of converting CO$_2$ into methanol in a process that resembles photosynthesis – a closed loop cycle in which every molecule of CO$_2$ released by burning methanol on board an automobile would be cancelled out by a similar molecule captured to make it. Olah, a Hungarian who fled his home country in 1956 and became a U.S. citizen, is co-inventor of the direct methanol fuel cell, which uses methanol to generate electricity, with the byproducts of CO$_2$ and water. While developing the concept, it occurred to him and his team that the process can be reversed. CO$_2$ from sources where it is present in high concentrations, like flue gases from a power plant burning natural gas could be combined with water, using electric power, to form methanol. Eventually, Olah believes we could just take CO$_2$ out of the air. Industry is beginning to express interest in this approach. In August 2008, the Japanese Mitsui Chemical Inc. announced its plan to construct a pilot plant for production of methanol from industrial CO$_2$ effluent and photocatalyst produced hydrogen.[7] The challenge is to chemically reduce carbon dioxide into carbon monoxide, a building block for fuel. This is not a simple process. CO$_2$ is a clingy molecule, and the carbon-oxygen bond requires a reliable catalyst to break it. If CO$_2$ is heated to 1,700°C, it splits into carbon monoxide and oxygen. At that point, the Fischer-Tropsch process can be used to convert the CO into hydrocarbons. But what source of energy could be used to generate such heat without emitting more CO$_2$ into the atmosphere? Olah puts his faith in nuclear power. Others believe that energy from concentrated solar power can do the trick. Researchers at Sandia National Laboratories in New Mexico, for example, use sunlight through a chamber containing mirrors to divide CO$_2$ into carbon monoxide and oxygen. The carbon monoxide is then synthesized with hydrogen to form almost any type of liquid fuel. According to Sandia, such solar chambers could provide enough fuel to power 100 percent of America's vehicles using 2,200 square miles, one tenth of California's sunny, yet almost unpopulated, Mojave Desert.

Green goo

Another interesting way to utilize CO_2 can be found in the lowest part of the food chain: algae. This slimy little plant requires large amounts of CO_2 to grow. Once harvested, algae's starches can be made into ethanol, while its lipids are refined to make biodiesel or jet fuel. The remaining protein can become livestock or fish feed or even co-fired at coal power plants. As one of the world's leading algae experts, Isaac Berzin explains "every stage of algae growth and production reduces carbon emissions and limits dependence on foreign oil, and each stage is profitable. Growing algae reduces carbon emissions. Developing algae-based biofuels reduces emissions from automobiles and reduces the amount of imported oil. Co-firing algae and coal reduces the amount of coal required to generate power, thereby reducing carbon emissions." Berzin teamed up with the utility Arizona Public Service Company (APS) on Redhawk, its natural gas power plant outside Phoenix. There, a 2,000 acre algae farm has been built which feeds on the power plant's CO_2 emissions, hence converting the plants' carbon dioxide emissions into a useful resource.

Algae is one of nature's miracles. It doubles its mass every few hours, and it produces 100 times as much fuel per acre as soybeans, currently the main crop for biodiesel production and 10 times more ethanol per acre as corn. The land requirement to replace all of the liquid fuel used in the United States with algae-derived fuel is estimated at roughly 15,000 square miles. This sounds like a lot, but to put that in perspective, consider that 15,000 square miles works out to roughly 2 percent of the land currently used for crop farming in the United States and that algae ponds can utilize land that is unsuitable for farming. Furthermore, the unique ability of algae to grow in saline water means that it can be grown in areas of the country in which saline groundwater supplies prevent any other useful application of water or land resources. But in order to use algae ponds as CO_2 sinks, they should be located in the vicinity of power plants, cement plants or any large plant with a significant and reliable stream of CO_2 exhaust. A former RAND analyst who examined the algae technology employed in Redhawk concluded that perhaps three-quarters of the plants have sufficient room for such a facility, and that if such algae-to-fuel facilities could be located at every operating coal plant in the U.S., it would not only be possible to produce about 13.5 billion gallons of biodiesel and 8.5 billion gallons of ethanol per year but also to eliminate about 500 million metric tons of CO_2. This represents more than 20 percent of carbon dioxide emissions from coal-fired power plants.[8] Algae could theoretically supply about half of the alternative fuels President Bush called for in his 2007 State of the Union Address (35 billion gallons by 2017), but do so while reducing the amount of land that would be needed by three-quarters while avoiding competition with food supply.

Approaches like these – and others – can turn CO_2 from a potentially problematic greenhouse gas into an essential raw material for an alcohol-based economy and at the same time alleviate the problem of our diminishing fossil fuel resources. None of this will happen overnight nor is it free of technological challenges, but, as described before, the sequestration alternative is not exactly a sure thing either. It would take some hubris on our part to determine now which, if any, of these approaches would ever become economically and technologically viable. But from an economic standpoint, they all enjoy a head start in comparison with sequestration simply because they treat CO_2 as an asset rather than a liability and are not burdened by the prerequisites of massive legislation and regulation that sequestration entails. Which is why we should overcome our myopic fixation with cherry picked solutions and various R&D favorites and instead we should explore the full range of solutions science enables us to deploy, and let the best of them win. In the near term, an Open Fuel Standard would guarantee a sizable market for CO_2-derived fuels by ensuring that cars will be able to run on them.

⌘ ⌘ ⌘

7

THINKING *INSIDE* THE BOX:
WHY BATTERIES ARE SO EXPENSIVE AND WHAT WE
CAN DO TO BRING DOWN THEIR COST

In the previous two chapters, we outlined the two pillars essential to breaking oil's monopoly in the transportation sector: the introduction of an Open Fuel Standard which would open the transportation fuel market to competition and the electrification of transportation through the deployment of PHEVs and pure EVs. While complementary, the two pillars are fundamentally different and raise challenges of different sorts. The main difference is in the ramp up effort. The alternative liquid fuel route is easy to implement on the vehicle side. Flex fuel technology is mature and low cost. For automakers to manufacture flex fuel cars would require no more than a tweak in the assembly line, and the cost of doing so is minimal. Within a few years, every car sold in America can be a flexible fuel vehicle. But while the cars are easily deployable, supplying them with fuel is not a trivial challenge. Refueling stations would have to retrofit their pumps – today only 1,300 of 170,000 refueling stations offer alternative fuels – and, most important, fuel supply would have to increase in spades. Hundreds of new alternative fuel plants will have to be sited, engineered and built, and a gigantic amount of feedstocks of various sorts will have to be grown, collected and mined. This massive undertaking will take years to complete. When it comes to electric transportation, the challenge is exactly the opposite: the fuel and the refueling infrastructure are readily available – most of us have access to the grid, and electricity, as described in Chapter 5, is abundant – but the vehicle ramp up is costly and slow. This reality often escapes even some of the most committed energy independence advocates. During his presidential campaign, President Barack Obama delivered an energy policy speech in Lansing, Michigan, in which he stated his goal of putting one million PHEVs on U.S. roads by 2015. He also promised that 50 percent of cars purchased for the federal fleet will be plug-ins by 2012. Finally, he pledged to convert the entire White House fleet to plug-ins, "as security permits," within one year of an Obama administration arriving at 1600 Pennsylvania."[1] Obama was not the first president to make such commitments. In January 2007, his predecessor, President Bush, issued an executive order mandating federal agencies buy PHEVs in order to "give some surety to those who have invested in new technologies to know that the federal government is going to be a purchaser, when commercially available."[2] Throughout the world, country after country and city after city have made pledges to go electric. In 2008-2009 alone, Japan, Australia, Israel, Portugal, Spain, Ireland, Denmark, Monaco, the United Kingdom and Canada announced ambitious plans to convert part of their automobile fleet to run on electricity. Israel, the first country to embrace the electric car revolution, signed an agreement with Project Better Place to build a recharging infrastructure throughout the country, slashing the tax rate on cars powered by electricity to encourage consumers to buy the vehicles once they are available.[3] With its motto "How can we make the world a better place? One electric car at a time," Better Place infected Denmark shortly thereafter with the EV bug. The Danish energy company DONG Energy announced it will establish an electric car network in the Scandinavian nation with about 20,000 recharging stations powered by wind. The state of Hawaii followed suit announcing its plan to seed its islands with 50,000-100,000 charging stations by 2012 in order to use electricity to displace gasoline in some portion of its 1.1 million vehicles.[4] The state of California announced plans for a $1 billion infrastructure in the San Francisco bay area to support a huge fleet of electric cars; and the city of Paris has plans to deploy 4,000 electric cars by the end of 2009. From the Clinton Global Initiative to the Google Foundation, the philanthropic world too has become plug-ins crazed. Google alone put $10 million into advocacy efforts and to promote research, design and development of plug-ins.[5] In 2007, the company launched the RechargeIT initiative in which Hybrid Toyota Priuses and Ford Escapes were converted to PHEVs and were added to Google's fleet.[6] Google's Clean Energy 2030 plan proposed a rapid ramp up of sales of PHEVs, starting with 100,000 in 2010, and increasing to 3.2 million annual vehicle sales in 2020 and 16.5 million in 2030. Seventy percent of these vehicles would be plug-in hybrids, with the remainder being all-electric vehicles.[7] And so it goes.

But despite the excitement, the electric car revolution is still like a vaccine for AIDS. Most of us want to see it happen, but the hurdles that need to be surmounted to turn the vision into reality are much higher than most of us would expect. Electric cars have been around for a century, and the history of their struggle is well described in popular works like Edwin Black's *Internal Combustion* and the movie *Who Killed the Electric Car?* The technology of electric motors is mature, efficient and reliable, and very few would doubt that electric cars are the natural successor to the internal combustion engine. Even those who still wish for hydrogen fuel cells should recall that in the end a fuel cell car *is* an electric car. The only difference is that in the former the energy is stored in a form of hydrogen while in the latter it is stored in a battery. With fewer moving parts and friction, the electric motor has 68 percent efficiency versus 40 percent for the internal combustion engine. For a demonstration of how durable and reliable electric motors are, visit the elevator shaft of any high rise building where electric elevators have been carrying thousands of tons of people and equipment up and down flawlessly every day for decades powered by similar motors. With fewer components, these motors are also considerably cheaper than the internal combustion engine. They are also lighter. The motor of the Tesla Roadster, perhaps the most powerful electric engine for vehicles, weighs only 130 pounds. Its battery pack, on the other hand, weighs 1,000 pounds. And it is in the battery where the problems lie.

The battery challenge

We have come a long way since 1886, when a prominent electrician conveyed the general sentiment in the battery field, saying, "The man who comes out with a kind word for the secondary (rechargeable) battery is set down either as a knave or a fool."[8] While significant progress has been made in the field, advanced batteries are still the barriers to mass market penetration of electric cars. And as was the case a century ago, the problems still come down to basic materials and construction. For electrons to replace oil, these problems will have to be resolved soon.

Batteries are boxes that convert chemical energy to electrical power and have the capacity to store that energy. The first thing to understand is how incredibly complex advanced auto batteries are in comparison to the primary (non-rechargeable) batteries we use in our watches and toys or even the secondary (rechargeable) batteries we use in power tools and cell phones and to start our cars. In order for the box to become an effective energy storage device capable of moving a heavy vehicle over long distances and able to charge and discharge over and over again thousands of times, it needs to be significantly "smarter" than the conventional cell. The batteries we use in our flashlights are used once and then discarded. Our flashlight then happily accommodates a new battery regardless of whether it's Duracell, Energizer or any other brand of alkaline battery as long as it's the right standardized size. Secondary batteries in cell phones and cameras are much less standardized and their interchangeability is limited. This is a disadvantage for the user, but it creates a captive aftermarket for the manufacturer. The lead-acid auto batteries which were invented in 1859 by French physicist Gaston Planté and which we use today in our cars are not nearly as smart as the ones used in PHEVs. All they are asked to do is power the starter motor, the lights, and the ignition system of the vehicle. As such, they require neither a control system nor high voltage. But when it comes to an advanced auto battery, the kind used in a PHEV and EV, the requirements are far more stringent than they are for portable consumer electronics or a standard lead-acid.

First, safety is a huge consideration. With today's gasoline and diesel powered engines, every year, more than 250,000 vehicle fires occur in the United States, resulting in about 490 deaths and 1,275 injuries.[9] That is one of the risks we take when we drive. Yet as they introduce new PHEVs and EVs to market, automakers can't risk even a rare serious battery-caused fire. The consequences for consumer confidence in EVs in case of a mishap could be incalculable. So a great deal of effort has been expended to tweak battery chemistry and design to ensure that the risk of technical malfunction causing a fire or explosion in a lithium ion (LiIon) battery, most automakers' battery of choice for electric transportation, is reduced to a bare minimum. Durability and performance are no less of a challenge. Unlike your laptop battery, which works several hours at a time mostly at room temperature, a car battery operates mainly outdoors at times in a very rugged unpleasant environment with significant variability in temperature, humidity, and other factors. We also expect our car battery to last longer than the one powering our cell phone. Most of us wouldn't consider keeping our Blackberries for more than 10 years, but when it comes to our car the durability requirements cited by most automakers as a condition for deployment of advanced batteries is 10 years of life or 150,000 miles whichever comes first.

Because of the early stage of the technology, there are no uniform standards for advanced auto batteries. Unlike primary and secondary batteries, advanced batteries communicate with the car's control system. Which means that advanced auto batteries must be designed to dialogue with a specific control system just like the two lobes of a human brain communicate with each other. Ford, GM, Nissan and the other automakers have each contracted with different battery manufacturers to develop their own dedicated packs each filled with cells containing unique anodes, cathodes and separators made from a variety of materials and using various chemistries. This means zero interchangeability between batteries and cars. A battery developed by NEC Global for Nissan would look nothing like the battery developed by Johnson Controls for Ford or that developed by LG Chem for GM. While the proliferation of different design approaches – may a thousand flowers bloom – has a variety of advantages in terms of increasing the likelihood of a homerun or even different homeruns for different market niches, the implications of this technological free-for-all are stark for companies like Project Better Place that focus on rapid battery swapping and thus require design uniformity across both battery and auto companies in order to succeed. Automakers are also diverse when it comes to the preferred battery chemistry. The hybrid vehicles today mass produced by Toyota, Honda, GM, Ford, and Chrysler utilize a nickel-metal-hydride (NiMH) battery. Pure electric vehicles like the Toyota RAV4-*EV* which has a 100+ mile range also use NiMH batteries. But many in the automotive world believe that NiMH has reached its zenith as the preferred battery chemistry. While longevity and reliability are advantages of the NiMH battery, its weight, cost and relatively poor efficiency have given prominence to LiIon. A LiIon battery is lighter and stores two to three times as much energy per pound as a NiMH. It also has a low self-discharge rate of approximately five percent per month, compared with over 30 percent for NiMH. Its low weight gives LiIon chemistry particular appeal. As it is, the Volt's battery pack weighs 400 pounds and Tesla's 1,000. With NiMH chemistry, those batteries' weights would have doubled, putting a huge burden on the engine. LiIon has another important advantage: it prefers partial discharge to deep discharge. In other words, the battery's durability is not compromised when we recharge it while partially charged. In fact, it prefers that we avoid taking it all the way down to zero. If the voltage of the cell drops below a certain level, it's ruined. Yet, despite the apparent advantages of LiIon technology, it remains to be seen whether it will indeed be the battery of the future.

A thin plastic sheet

In order for PHEVs to become commercially appealing, we must understand what it would take to cut their cost to a level at which they can compete with conventional cars. Let's dive into the guts of the LiIon battery. The groundbreaking research by a group led by a physicist at the University of Texas in Austin led to the introduction of the first commercial LiIon battery by Sony in 1991. That lithium cobalt oxide ($LiCoO_2$) battery revolutionized portable electronics. An automotive battery pack is made of cells that resemble small laptop batteries. Each of those cylinders contains a positive electrode (cathode) normally made of lithium-transition-metal oxides, such as $LiCoO_2$, but some manufacturers use iron nanophosphate or manganese spinel. The negative electrode (anode) is commonly made from graphite or lithium-intercalated carbon (LiC_6). Current is generated as the lithium ions flow through an electrolyte, such as a lithium salt dissolved in an organic solvent, from the positive electrode to the negative electrode and attach to the carbon. During discharge, the lithium ions move back to the anode from the cathode. The anodes and cathodes are relatively easily produced, and as the next chapter shows the raw materials needed for their production, lithium carbonate being the most important one, are abundant and cheap. In 2008, the price of one kilogram of lithium carbonate was $6. The lithium carbonate requirement of 1kWh is roughly $0.6-$1.4/kg depending on the battery type. This means that the cost of lithium for a 16kWh Volt battery is no more than $130. The cost of the other metals is equally negligible. In 1996, the aforementioned Texas professor's team discovered a chemistry that brought LiIon battery safety and durability to a whole new level, and, perhaps more importantly posed great promise in terms of lowering cost at mass production. The lithium iron phosphate ($LiFePO_4$) battery differed from the batteries used in laptops and cellphones with its higher resistance to thermal runaway – and thus much lower likelihood of fires. It also utilized iron and phosphate, materials that are lower cost than cobalt and significantly more available.

Surprisingly, what make LiIon cells so expensive are hair thin sheets of plastic called separators. As the name implies, separators separate the positive and negative electrodes while allowing ions to pass through. Their role is to prevent catastrophic battery failures that have led in the past to flaming laptops and the recall of millions of batteries. They are made of polyolefins, the same plastic material that is used, alone or in blends, in sportswear, undergarments, outdoor furniture, rugs, filters, and electrical insulation. In fact, automobile seat covers contain more polyolefins than the battery. But battery separators are much 'smarter' than car seats. They are made of several super thin sheets of porous membranes that allow the flow of ions. The more increased the porosity, the greater the flow of ions and hence the more power the battery can exert. For safety purposes, at least one of the membranes is so heat sensitive that it closes its pores when the battery cells reach a certain temperature. This type of technological marvel and the intellectual property and research efforts that go into its development is the reason battery separators are perhaps the world's most expensive plastic sheet. Their cost per battery is roughly $1,000-$1,800 per PHEV and $2,000-$2,100 per EV. So if we are to bring down the battery cost, we must find ways to mass produce low cost, yet equally smart and safe, separators. Who makes those fancy separators? Until recently, roughly 50 percent of the world's separator production capacity was owned by the Tokyo-based Asahi Kasei Corporation, a $10 billion in sales manufacturer of chemicals, fibers and related consumer products. The rest is produced by a handful of other companies. But in May 2007, one of the world's biggest companies surprised the world, unveiling an advanced performance separator with enhanced permeability and higher meltdown temperature, that didn't compromise the shutdown temperature and mechanical strength. The company's name was ... Exxon Mobile.[10]

This thin, polyethylene-based porous sheet is the most expensive component
of a Li-Ion battery
Source: Tonen Chemical Corporation

A prize to beat Exxon

In conjunction with its affiliate, the privately owned Tonen Chemical Corporation, the oil giant has raised the battery safety bar higher than any other competitor. It is Exxon that works to keep our batteries from exploding. This is probably the point where cynics would suggest that Exxon's interest in critical components of car batteries stems from its desire to perpetuate oil's monopoly in the transportation sector, that they will sit on the patent, drive the battery price up and keep us addicted to oil forever. The evidence for such conspiratorial thinking is flimsy. Oil companies have shown growing interest in technologies that can displace oil. Royal Dutch Shell owns 50 percent of Iogen Energy Corporation, one of the world's leading cellulosic ethanol producers. Chevron owns half of Cobasys, a major developer of NiMH battery packs for hybrid vehicles including GM's Saturn Vue Green and Saturn Aura. And the Israel Corporation, which owns the Israeli refining industry, is the leading investor in Project Better Place and a number of biodiesel companies. Part of the rationale for these investments might have to do with PR. Investment in clean technologies works to improve those companies' public images as so-called enemies of the planet. The reality is far less cynical. Despite handsome profits in recent years, the companies most people think of as Big Oil are in distress. Since 80 percent of the world's oil reserves are owned by government-controlled national oil companies (NOCs), international oil companies (IOCs) find it increasingly difficult to get access to new oil to replace their depleting reserves. More than anyone else, they are fully aware of the fact that, as Chevron CEO David O'Reilly put it, "the era of easy oil is over." For IOCs, investment in alternative energy is an insurance policy against an uncertain future. In many cases such investments also prove highly lucrative. This was the case of Exxon's flirtation with battery separators. As part of its diversification strategy, Exxon has been involved for decades in the chemical industry through its subsidiary Exxon Chemicals. Its interest in battery separators emerged as early as the 1970s, well before automotive LiIon batteries were on anybody's mind. In fact, it was American chemist Manley Stanley Whittingham, an Exxon scientist, who in the Seventies proposed the very earliest iteration of lithium-ion battery chemistry. In 1991, Exxon's separator film was already used in the first commercially produced LiIon batteries for Sony's cell phones. So when it comes to Big Oil bashing conspiracy theories, count us as skeptics. This doesn't change the fact that while for Exxon the separator business is just another profitable business, for the rest of us it's about the future of our civilization. One would hope that at mass production, Exxon would quickly recover the cost of its intellectual property and the price of the separators would drop significantly. But waiting for this to happen could mean years of unnecessary delays on the path to an electric future. One way to speed the development of lower cost separators is to adopt a limited version of the "McCain prize doctrine" to encourage the invention of competitive membranes. In his 2008 presidential campaign, Senator John McCain called for a $300 million prize to whoever can develop a battery that will "leapfrog" the abilities of current hybrid and electric cars. He said the new automobile battery should have "the size, capacity, cost and power to leapfrog the commercially available plug-in hybrids or electric cars."[11]

Throughout history, cash prizes have been an effective mechanism to advance innovation and human achievement. It was the $25,000 Orteig Prize that enticed Charles Lindburg to fly the *Spirit of St. Louis* across the Atlantic and the prestigious $10 million Ansari X Prize that inspired American aerospace engineer Burt Rutan and software magnet Paul Allen to develop SpaceShipOne, the spaceplane that on June 21, 2004 completed the first privately funded human spaceflight. Less known is the story of the prize that broke salt's long-held monopoly over food preservation. In 1800, Napoleon Bonaparte offered a 12,000 franc cash prize to anyone who could devise a practical method for food preservation for armies on the march. Malnutrition was a major problem for the 18th century French Army and a major operational problem for Napoleon who was planning his Russian campaign. A French confectioner and former chef named Nicholas Appert responded to the challenge. After years of experimentation, Appert found a way to preserve food in glass containers. He won the prize in 1810. That same year, fellow Frenchman Pierre Durand patented his own method, only this time using a tin receptacle, thus creating the modern day process of canning foods. In the same vein, a prize for the developer of low-cost separators – as well as other components of the battery supply chain – would be an effective way to jumpstart innovation. All that is needed is to define the right targets for cost, durability, safety and performance, declare a handsome bounty and entice an army of chemists and inventors to come up with a better product than Exxon's.

The good enough

Currently, LiIon technology presents a tradeoff between cost and battery durability. The longer life the battery has both in terms of years and in terms of number of charge and discharge cycles, the more resilient it is in the face of weather variability and so forth, the more expensive it generally is. Auto companies selling to the U.S. market work under the assumption that the average consumer expects batteries to last 10 years or 150,000 miles, and are reluctant to introduce cars to the market in which the battery may need to be replaced sooner than that. Therefore, the battery cost they face is quite high. Depending on the size, a LiIon battery pack for PHEV that meets these durability requirements costs $8,000-$15,000 (roughly $1,000/kWh.) A 30kWh EV pack costs about $18,000 ($600kWh). To prevent such cars from becoming money losers, their battery cost must come down to under $300/kWh.[12] "Clearly we never made any secret out of the fact we're not going to make much money off the Volt until the cost of batteries go down," said Bob Lutz, GM's vice chairman for global product development.[13] Chinese companies, on the other hand, are much more comfortable with introducing PHEVs with battery characteristics and separators that are "good enough" yet affordable, taking the risk of having to replace some batteries earlier than desirable, and improving the technology as they go along. Just as in other industries like semiconductors, good enough products were introduced as soon they were ready and improved with each successive technology generation. Thus, Chinese-manufactured lithium iron phosphate batteries cost just half of their counterparts in Japan and the West.[14] Since the basic vehicle platform of Chinese auto companies is also extremely low cost, combining the pieces, they are able to introduce plug-in hybrids to the market at a cost of just over $20,000.

Remember the Austin professor who led the team that discovered the lithium iron phosphate battery? Ironically, his name is John B. Goodenough.

Whither Washington?

Embarrassingly, when it comes to advanced LiIon batteries, the United States is almost a non-actor. In 2008, Japan produced 39 percent of the world's supply, China 36 percent, South Korea 20 percent and 5 percent was shared by the rest of the world, U.S. included. This is a badge of shame to a country that aspires to lead the electric car revolution. To become a world leader in advanced battery development and production, a much more serious commitment would be needed. First, the federal government can play a critical role in helping to accelerate R&D in advanced batteries. The 2007 Energy Independence and Security Act (EISA) authorized[15] more than $500 million for PHEV, EVs and other related components of the electric transportation program.[16] At first sight, this may look like a lot of money, and many of the bill's authorizations, had they been funded with appropriations, would have contributed greatly to the electrification of transportation including demonstration programs, applied research and the creation of no fewer than four energy storage centers managed by the Department of Energy's Office of Science. Even so, a deeper scrutiny reveals that only a tiny part would actually address the challenge discussed in this chapter of bringing down the cost of batteries. On this front, EISA authorized a mere $50 million a year – equivalent to 100 minutes of imported oil at $60 a barrel – for ten years for basic research on batteries, hardly enough to make the United States a battery superpower. Ironically, it was the economic downturn of 2008 that prodded Congress to give a shot in the arm to America's efforts to electrify its fleet. As part of the American Recovery and Reinvestment Act that the U.S. Congress passed in early 2009 – the so called "Stimulus" – some $2 billion was allocated for funding Department of Energy grants to manufacturers of advanced batteries and battery systems.

We need to recognize – just like Japan did – that advanced battery technology is an issue of national survival. When such recognition comes, the rest follows. With such a mindset, the Japanese government is funding LiIon battery manufacturing research programs along with materials and systems development. Japan already supplies more than 70 percent of the NiMH batteries worldwide and it intends to maintain 60 percent to 70 percent of the global production of LiIon batteries. We must craft a comprehensive national battery strategy that systematically analyzes the entire supply chain of battery components, carefully identifies the bottlenecks and provides solutions to open them. A Sematech-like program focused on developing a manufacturing capability might help jump-start a homegrown battery manufacturing industry in the United States. Headquartered in Austin, Sematech (SEmiconductor MAnufacturing TECHnology) was formed 20 years ago as a partnership between the U.S. government and leading American semiconductor companies. Just like with advanced batteries today, in the mid-1980s the global semiconductor industry was dominated by Japan, and American manufacturers were struggling to regain a competitive edge. The only way to do so was by pooling resources in order to solve common manufacturing problems. Because semiconductors were perceived as a national security interest, the private sector's funds were met with significant grants from the U.S. Department of Defense. Over the years, the U.S. industry gained enough strength, and government support ended in 1996. Since then, Sematech has become an international consortium, and half of its members today are non-American corporations.[17] A similar pathway can be adopted for car batteries. The national security rationale is difficult to contest. Finally, result-oriented tax incentives to both producers and purchasers as well as procurement programs for federal, state, and municipal fleets could go a long way in developing a market for advanced batteries. The law passed by Congress in 2009 awarding the first 500,000 early adopters of PHEVs with a $7,500-$15,000 tax credit keyed to the battery size, which correlates with the capacity for oil displacement, is a great way to get over the early adopter hump.

As the backdrop of all of the above activities, it is important to remain open minded about other battery chemistries and realize that advanced battery technology is only at the beginning of its upward trajectory. While significant progress is made toward mass production of advanced LiIon batteries, it would be imprudent for the government to crown one battery chemistry "the winner," particularly in light of its questionable track record of picking R&D winners. With all their potential, the race is not over. In fact, it is only beginning. In this, it's worth paying attention to Jack Lifton, a veteran observer of the battery and technology metals world, who observed:

> The popular press, which due to the poor education of most of its practitioners, is the literal slave of the marketing and public relations departments of companies that stand to benefit financially from the public acceptance of one battery technology over another by means of a popularity contest in which the winner gets government subsidies to continue the development of his chosen technology, no matter what its actual or future value, has chosen to hyperventilate over the immature and untested lithium-ion battery technology alone. Ignorant or interest-conflicted spokesmen for any of the 26 lithium battery chemistries now being researched will say things such as "nickel-metal hydride battery technology is 'primitive'" compared to the undisclosed lithium ion technology they are offering to sell whenever it is developed. This type of statement is simply stupid.[18]

We don't know whether or not Lifton is right. But if he is, putting all of our weight behind the LiIon battery may be a strategic mistake. Battery technology is constantly evolving, and there are already several other well-understood battery chemistries, sodium nickel chloride and ZincAir are two examples that come to mind, and fuel cells (in vehicles fueled by methanol, not hydrogen) which also hold potential. In the future, we might even see a "hybrid battery" which combines two or more chemistries in the same pack.[19] Driving has three distinct power requirements: initial acceleration which requires short but strong power, normal acceleration, which requires less power, and cruise, which requires steady low power. In a hybrid battery, two chemistries can work simultaneously. LiIon, which can discharge very quickly, will fill the first category, making a small part of the battery pack. And other non-lithium compounds will fill the rest. A smart car controller will have to know the power discharge profile of the battery pack and request the proper part of the battery part to discharge.

The point is again choice. Choice in batteries, choice in chemistries and choice in raw materials. Of all these, our choice in raw materials could more than anything impact the geopolitics of energy in the world in the 21st century. Here is why.

⌘ ⌘ ⌘

8
BACK TO THE SALT BRINES:
TOMMOROW'S DEPENDENCIES

On September 10th, 2008, Philip Goldberg, U.S. Ambassador in La Paz, Bolivia's capital, got a message no ambassador wants to receive. He was declared persona non grata in his host country for his alleged "support of the opposition and encouraging the division of the country." In the midst of a wave of riots, the country's president, Evo Morales, an avid socialist and a protégé of Hugo Chavez, was edgy. His concern for the stability of his regime and his paranoia about U.S. meddling in his domestic affairs caused impulse to trump sound diplomatic judgment. Ambassador Goldberg was sent packing. The Bush Administration's retaliatory response was swift and decisive: "In response to unwarranted actions and in accordance with the Vienna Convention [on diplomatic protocol], we have officially informed the government of Bolivia of our decision to declare [the Bolivian] Ambassador Gustavo Guzman persona non grata," State Department spokesman Sean McCormack said. Several days later, President Bush asked to punish Bolivia economically, requesting withdrawal of Bolivia's trading concessions as part of the Andean Trade Promotion and Drug Eradication Act. (The U.S. is Bolivia's second-largest export market after Brazil, purchasing 15 percent of Bolivia's exports.) This out-of-the-blue, full blown, diplomatic crisis received almost no attention in the United States. Between the other big news items of that week, the collapse of Lehman Brothers and the AIG bailout, the latest headline from the presidential elections and the dramatic advance of Hurricane Ike toward Houston, Americans had little attention span even for a sacred event like the seventh anniversary of 9/11, not the least for a diplomatic squabble with a poor and relatively little known Andean country.

But what if instead of Bolivia it was Saudi Arabia that decided to sever its diplomatic ties with the United States? Of course, this would have sent shockwaves throughout every part of our society. After all, Saudi Arabia has oil, and oil underlies the global economy. Fast-forward two or three decades from now and a crisis with a country like Bolivia could have profound implications for America's wellbeing. Bolivia may not have oil, but it is the world's largest reserve of lithium, a critical component of the LiIon battery. If all of our cars are to be powered by LiIon batteries, then those who own lithium would gain enormous geopolitical influence. So while Saudi Arabia is central to our energy security *today*, Bolivia may enjoy a similar position *tomorrow*. "The green-car revolution could make lithium one of the planet's most strategic commodities," *Forbes Magazine* observed in November 2008.[1] Which reminds us that humanity's pursuit of strategic commodities and the various challenges associated with such dependencies will not end with the transition from oil. There will be new commodities to compete over, new players, new regions and new geopolitics. Bolivia might be the Saudi Arabia of the next oil, and this would no doubt create non-trivial challenges for U.S. foreign policy. But it doesn't have to be the case if we begin now, as we move to address our current energy dependency, to think about the next set of challenges that the post-petroleum age has in store for us.

In search of the lightest metal

The lithium used in LiIon batteries comes in the form of a chemical compound called lithium carbonate (Li_2CO_3), the same stuff used to treat manic and depressive disorders. To make all those millions of LiIon batteries, we need Lithium metal. A lot of it. For some perspective on how much lithium will be needed to power the world vehicle fleet, consider this: Our Blackberries require a mere 3 grams of Li_2CO_3. The lithium ion battery that powers a laptop contains 30 grams. A standard hybrid car with 1.5kWh LiIon would require roughly 1 kilogram. Turn this hybrid into a PHEV with a 10kWh battery and you need 6-10 kilograms of Li_2CO_3 depending on the technology. In order to go 40 miles on electric power, GM's Chevy Volt will have a 16 kWh battery which requires something in the order of 22 kilograms of lithium carbonate. Each one of the long range pure electric cars by Renault/Nissan Project Better Place is counting on will need about 30-45 kilograms, and the Tesla Roadster, the all electric sports car which can go up to 200 miles per charge will require more than 80 kilograms, more than is currently used to manufacture 2,500 laptops. To put it slightly differently, if all the lithium carbonate used to make the 78 million laptops shipped globally in 2007 were diverted into production of Teslas, it would be sufficient to make no more than 30,000 cars. This would not have been a problem if batteries were the only end user for lithium. But they aren't. Only 14 percent of lithium carbonate production – in 2007, 11,000 tons out of a total production of 81,000 tons – is utilized by the battery industry, mostly for power tools, cell phones and computers. The rest is used for the production of ceramics and glass, lubricating greases, pharmaceuticals and polymers, air conditioning and aluminum alloys, among other purposes. Clearly, if LiIon batteries become the batteries of choice for automotive use, the lithium requirement of the battery sector will grow in spades.

Consider this: according to the Energy Information Administration 2008 Outlook, by 2030, 38 percent of new cars will be hybrids, a projection that many find to be overly conservative. This equates to over 8,000,000 hybrids per year in the United States alone. If those hybrids are equipped with LiIon batteries, this market alone would require 8,000 tons of lithium carbonate, nearly one tenth of current global production. The lithium requirement of President Obama's plan to deploy one million PHEVs by 2015 is equivalent to the current needs of the global battery market for portable electronics. Roughly one third of the current total world production of lithium will be needed to meet the lithium requirements of the electric vehicle program announced by the Governor of Hawaii and Project Better Place. If this looks like a tall order, take a look at Google's plan to reduce fossil fuel consumption. Google's plan foresees a rapid ramp up of U.S. sales of PHEVs and EVs reaching 16.5 million units per year in 2030. Seventy percent of these vehicles would be PHEVs, with the remainder being EVs. Not a word from Google though on how to obtain the 190-280,000 tons of lithium carbonate needed to make those 16.5 million batteries. And this is without even considering the projected growth of 25 percent per year in lithium requirements for the portable electronics market and the vehicle electrification needs of other major auto markets like China, Japan and Europe. Get the picture? Clearly, if we are to peg our hopes to the LiIon battery, we must begin to assess lithium availability, the existing reserve base and, most important, the ways to ramp up production – and quickly.

The good news is that there is plenty of lithium to go around. Lithium is the 28th most abundant element in nature. One of the most quoted studies into material availability for a future EV fleet was carried out by Bjorn Andersson and Inge Rade of Chalmers University of Technology in Sweden which shows that there is sufficient lithium in the Earth's crust to power anywhere between 200 million to 1.2 billion EVs with LiIon Manganese based batteries.[2] The world's leading authority on lithium, veteran geologist R. Keith Evans, claims a total of 28.5 million tons of lithium, equivalent to nearly 150 million tons of lithium carbonate of which nearly 14 million tons lithium (about 74 million tons of carbonate) are at active or proposed operations.[3]

Lithium metal can be obtained through traditional hard-rock mining of ores called pegmatites which contain the lithium bearing silicate spudomene. This traditional mining is time, energy and cost intensive. A more efficient and cost effective way to get lithium is from lithium rich brine deposits that occur in closed basins in high evaporation, arid environments. The process is quite similar to the production of salt. The brines are pumped to a series of evaporation ponds where the lithium chloride solution is allowed concentrate, soda ash is then added, and the end result is lithium carbonate that is then purified, dried and shipped. Currently, 36 percent of the world's lithium comes from Salar (in Spanish "salt lake") de Atacama, a lake 700 miles north of Santiago, Chile. Other major producers are the Salar Hombre Muerto in Argentina, and a few locations in Tibet, China and Brazil. But where is Bolivia? More than 30 percent of the world's reserve base of lithium brine is located in its Salar de Uyuni, the world's largest salt flat. About the size of Connecticut, the Salar is located in southwest Bolivia, near the crest of the Andes. It is home to 10 billion tons of salt, and were salt still a strategic commodity, the Salar would be a name known to every third grader. Provided that the LiIon chemistry gains dominance, not too many years from now this magnificent salt desert could become as central to the world's economy as the world's largest oil field, Ghawar, in Saudi Arabia. This despite the fact that the quality of its lithium is inferior to that of the Salar de Atacama. But despite its 40 percent share, today Bolivia is a non player in the world's lithium market. In fact, its lithium production is almost zero. (Imagine Saudi Arabia and Iran, cumulatively sitting on top of 35 percent of the world's oil reserves, not producing a drop of oil.) The Andean nation of eight million is the poorest country in Latin America – over 70 percent of Bolivians live below the poverty line – so lithium production could become a tremendous economic engine. But Bolivia is not a country known to be friendly to foreign industry. It had a long history of foreign exploitation, beginning with Spanish colonists sending home its silver and leaving the Bolivians deprived. Millions of Bolivian Indians and slaves died extracting silver and tin for the Spanish. Bolivians are concerned that, as before, the money from their lithium will end up in the pockets of the elites rather than benefiting the population at large. The Bolivian government is therefore not quick to tap into its treasure of 5.4 million tons of lithium reserves. "We want to send a message to the industrialized countries and their companies," said Bolivia's minister for mining, Luis Alberto Echazu, "We will not repeat the historical experience since the fifteenth century: raw materials exported for the industrialization of the west that has left us poor."[4] President Morales announced: "The state doesn't see ever losing sovereignty over the lithium. Whoever wants to invest in it should be assured that the state must have control of 60 percent of the earnings."[5]

Salar de Uyuni, Bolivia

U.S. relations with Bolivia are complex. Despite the strong anti-American sentiment by the current regime in La Paz, the United States still provides foreign aid to Bolivia. In 2008 alone, USAID provided Bolivia with $85 million in development grants. But tension over the country's role as major coca grower and key ally of Venezuela's president Hugo Chavez and his "anti-imperialist block" is likely to hinder U.S. attempts to participate in the country's lithium industry. To change the trajectory, the Obama Administration would have to assign U.S.-Bolivia relations a special status, emphasizing the need for the country's equitable and sustainable development and ensuring that unlike in previous historical experiences, the country's natural resources would directly benefit the people. The good news is that even without Bolivia's lithium, and contrary to pessimistic projections by some analysts prophesying "peak lithium," Chile, Argentina,

Australia, Russia, Zaire and China could provide ample supply for many years to come, provided sufficient investment is made. The bad news is that the United States, a holder of a non-trivial lithium reserve has very limited domestic production capacity. Two mines in North Carolina were closed, one in 1986 and one in 1998, and the only active lithium production plant remaining in the United States is a brine operation in Nevada. (The exact magnitude of the company's production is not disclosed to the public.[6])

Cobalt and Rare Earth Elements

Lithium is not the only commodity that could become strategic as we move toward electric transportation. State-of-the-art cathode materials contain metals such as cobalt, manganese and nickel, while electric motors require magnets made from neodymium, a rare earth metal which is also used for electric generators, computer hard drives, lasers, lamps and wind turbines. The NiMH battery of one Toyota Prius requires approximately 1.5 kilograms of cobalt and 12 kilograms of the various rare-earth oxides. The electric motor alone requires 1-2 kilograms of neodymium. PHEVs and EVs would require a similar amount of neodymium as their electric motor is basically similar, but due to a larger battery a much higher content of cobalt, up to 5 kilograms, would be required.[7] In 2007, global cobalt supply stood at 62,000 metric tons, most of it coming from the war-torn Democratic Republic of Congo, Zambia, Australia, Canada, Tibet, Siberia and Cuba (global reserves are estimated at 7,000,000 tons.) The United States doesn't mine cobalt. Cobalt is generally a byproduct of nickel and copper production, so whenever trouble arises in nickel and copper mines, as in the case of the violent civil war in the Shaba region in Congo, the supply of cobalt is also affected. According to Irving Mintzer from the Potomac Energy Fund, unless cobalt content in advanced rechargeable auto batteries declines significantly and assuming demand by other end users remains unchanged, global cobalt production would have to more than triple by 2030 to meet the worldwide demand created by the deployment of 24 million standard hybrids, PHEVs and EVs.[8]

The situation in the market for rare earth elements is even more complex than the conditions in the cobalt market. Rare earth elements are a family of metals on the periodic table, critical raw materials for hundreds of applications from consumer electronics to precision-guided weapons. Called "rare" for historic reasons, most of these elements are actually quite abundant in the earth's crust. Production capacity is another matter. Let's start with neodymium. In 2006, global neodymium demand was 7,300 metric tons. Depending on the demand scenario, and assuming all other end users for neodymium remain frozen, world demand under the high scenario would increase by a factor of five.[9] Highly ambitious plans to dramatically increase wind power capacity such as the Pickens Plan – significant quantities of neodymium are required for the production of the permanent magnet generator that goes on top of one wind turbine – could translate into competition between wind turbines and electric motors for the same resource (not to mention already existing end uses such as computer hard drives and headphones used with iPods and MP3 players.)

Want more of the rechargeable NiMH batteries used in hybrid cars like the Prius and some electric bikes? Can't make them without lanthanum. Just like the other rare earth elements, lanthanum production is complex. The ore – which is comprised of several different elements – is mined, crushed and milled, and then oxidized. There is only *one* country in the world today that can take the oxide and convert it to metals, and that's China. Due to a gradual shutdown of processing operations of rare earth oxides in the United States and Japan, once pioneers in the field, today China mines and processes almost all of the world's rare earth elements, lanthanum included. The world's largest deposit for rare earth is the Baiyunebo deposit located in Baotou, Inner Mongolia, China. In 1992, Chinese President Deng Xiaoping pointed out, "There is oil in the Middle East; there is rare earth in China." In recent years, China has done a laudable job ramping up production of rare earths, but due to its growing domestic demand its ability to export the metals is falling. Millions of electric motorbikes are built and sold in China each year, and increasingly such vehicles utilize NiMH batteries rather than lead-acid types. As a result of its growing domestic needs, China exports less and less to the rest of the world. In January 2009, the Chinese Ministry of Commerce reduced the amount of rare earth elements that can be exported by 34 percent.[10] Experts predict that the Chinese will be internally consuming many of those rare earths, if not all of them, by about 2013.

So if we are to have more NiMH battery-powered hybrid cars and electric motors, we will have to ramp up production of rare earth elements in countries other than China. Outside of China, there are substantial reserves of rare earth minerals in Australia, Russia, South Africa, Brazil and India. The United States also has considerable reserves – 14 percent of the world's total. But as with other critical materials, we produce zero. As with lithium and other critical metals, today, we prefer to be totally dependent on foreign suppliers of those materials necessary for our energy future.[11] (While a separate issue from the oil alternatives effort, it is worth noting that the above also applies to other technology metals and rare earth elements necessary for the green energy revolution desired by many. Want billions of those compact energy-efficient fluorescent light bulbs? Try to make them without europium, cerium, terbium and yttrium. Thin-film photovoltaic solar panels? Better check on the status of indium, gallium, selenium and tellurium.)

Addressing the material constraints

One way to stretch critical metals is through battery recycling. When batteries reach their end of life, the recovered metals can be used to produce new ones. Such a closed-loop recycling system with end-of-life management guarantees the best utilization of the world's metals. But battery recycling, particularly of LiIon, is still a nascent industry. Today, only 20 percent of total portable batteries in the world are being recycled. The figures for the U.S. are even lower. The Rechargeable Battery Recycling Corporation, (www.rbrc.org) a non-profit group formed by the rechargeable power industry and supported by companies like Bosch, Lowes, Best Buy, RadioShack and Office Depot, is playing an important role in promoting a recycling program for portable electronics. Since the metal content in car batteries is much higher than in portable electronics such programs need to be established for them. In industrial countries like the U.S., the environmental and occupational health regulatory cost of operating battery recycling facilities is ever-increasing, and the prices offered for the recycled materials are often too low. The economics of recycling are, of course, a function of the price of the recycled metal. Typically, old batteries are shipped to developing countries where they are recycled under minimal environmental standards. This may cause a slew of severe environmental and health problems for the populations of these countries, but it enables the best economic use of technology metals.

Another way to address the materials challenge is to extend the life of the battery. Longer battery life means fewer batteries needed, less lithium, less cobalt, less mining, less pollution, less demand for foreign metals. LiIon batteries last longer if strained less. High current discharge and recharge during acceleration and braking shorten the life of the battery. The battery can be buffered from these stresses with high power density devices called ultracapacitors. Ultracapacitors store electrical charge for later use at much higher density than in batteries. Just like other advances in electric storage, they too were developed in part by Big Oil. Today's ultracapacitors originated with the work of Standard Oil of Ohio Research Center (SOHIO) in the early 1960s. Over the years, the technology advanced and was applied to fuel cell vehicles. Unlike rechargeable batteries which begin to wear after several hundred charge-discharge cycles, ultracapacitors can endure millions of cycles and as a result their lifespan is much longer. Since they don't have the reactive chemical electrolytes found in rechargeable batteries, they do not present disposal and safety hazards. AFS Trinity, a privately-owned company headquartered in Bellevue, Washington, was the first to demonstrate a drive train for a PHEV with a two-part energy storage system which combines LiIon batteries with ultracapacitors. Such a storage system gets the best of both worlds: the lightweight and high energy density of LiIon batteries and the small size and high power density of ultracapacitors. AFS Trinity's Extreme Hybrid™ vehicle achieves top speeds and rapid acceleration in electric-only mode equal to a conventional hybrid while its design allows for a smaller internal combustion engine.

But in the end, if we are serious about the electrification of transportation, in addition to the above measures we have to mine – and a lot. The electric revolution will not come through words and hype but through a systematic treatment of all the components of the car and battery supply and manufacturing chain – from mine (or brine) to wheel. Ensuring future supply of critical metals will require a joint effort by government, industry and relevant constituencies. For example, environmentalists who still lament the killing of the electric car and who promote the so-called "green economy" should realize that the road to electric cars, efficient light bulbs, and both solar and wind energy production passes through the mine and entails easing some of our mining laws and opening new areas for exploration and recovery. To put it plainly: one cannot go green and tout energy independence while reflexively opposing any mining activity and lobbying for the designation of millions of acres where essential minerals can be found as wilderness. The real inconvenient truth is that more renewable energy entails easing some of our mining laws, reducing mining taxes, alleviating heavy-handed environmental regulations and opening new areas for exploration and recovery. No industrial nation can be totally self-sufficient, but if we are to be dependent on imported raw material for our energy future we must ensure that this dependence is carefully managed and that no single country can control the supply chain of industries critical to our existence. Like with oil, we must have stockpiles of essential materials sufficient for times of emergency, while our foreign policy establishment should begin to consider how our future energy needs might impact our foreign relations particularly since – as with oil – so many of the specialized materials are concentrated in unstable and/or potentially adversarial countries like Bolivia, Congo, Russia and China. The energy challenge we face cannot be addressed through faith-based energy policy but only through one that systematically addresses the entire supply chain of each component of our energy future, carefully identifies the bottlenecks and provides solutions to open them. Without such a comprehensive approach, our march toward energy independence could come to a grinding halt.

⌘   ⌘   ⌘

EPILOGUE:
THE CHOICE TO HAVE CHOICE

None of the challenges described in previous chapters are insurmountable and none should deter us from moving forward with the effort to break oil's monopoly over the global transportation sector. The rise of radical Islam and the prospects of a nuclear Middle East are enough of a reason to accelerate our efforts. The ultimate platform for fuel competition, the flex-fuel-PHEV is the best mechanism available today for shifting geopolitical power away from the Axis of Diesel. It would position South America and Africa on their vast lithium and sugar cane products in direct competition with Persian Gulf oil, China's coal and America's corn. Over the years, different global events could temporarily affect the supply of one fuel or another. Droughts, disease, frost or heat waves could impact biofuels production, tilting the balance to other fuel choices and feedstocks. Sometimes, during electricity blackouts, we'll thank our stars that our cars have a liquid fuel backup. We are likely to see the universe of batteries exploding with innovation, enabling more and more miles to be traveled on electrons. We've witnessed exponential progress in other industries. The Commodore 64, the best-selling single personal computer model of all time, was launched in 1982 with 64 kilobytes of RAM. Two decades later, most personal computers are sold with memories 10,000 times larger. It is not because more silicon went into their production but because data storage technology dramatically improved. If the same pace of progress will be achieved in energy storage, OPEC's days are numbered.

And while all this is taking place, enactment of the Open Fuel Standard is urgently needed so that the cars put on the road today are fully equipped to face the oil crisis of tomorrow. It seems so obvious, yet Congress has its own logic and priorities. Some needs of the American people get more attention than energy security. One of these is television. On June 12, 2009, America transitioned from analog to digital television broadcasts, ushering in what could be described as an open standard for television. This means that consumers now have a choice between buying a digital set or signing up to cable or satellite service and keeping their old antenna by installing a signal-dumbing converter box which allows them to get an analog signal. Without the converter, an analog TV shows snow on the screen. Regardless of whether the shift is a good idea or not – it probably is as it allows better usage of spectrum – it is a sad commentary on our priorities as a society. Strategic as television may seem to our illustrious leaders, it is not nearly as important as transportation. Yet, the same Congress that mandates choice in television reception modes denies us choice in transportation fuels. Choice at the pump is neither more difficult nor more costly to achieve than choice on the screen. To convert our televisions, Congress has already allocated nearly $2 billion in taxpayer dollars to provide every household $80 in coupons to subsidize the cost of the conversion boxes. To make a new car flex fuel costs an automaker no more than $100. Across 10 million new cars, that adds up to about $1 billion dollars per year.

Sad, but true: where we decide to put our money as a society reflects who we are, our priorities and how we view our future. In the early 2nd century, the Roman poet Juvenal, lamenting the decline of the Roman Empire satirized his political class in words that sound all too relevant to our current time of national decline: "for the People who once upon a time handed out military command, high civil office, legions – everything, now restrains itself and anxiously hopes for just two things: bread and circuses." Centuries later, during World War II, Winston Churchill commented: "Americans' national psychology is such that the bigger the Idea, the more wholeheartedly and obstinately do they throw themselves into making it a success. It is an admirable characteristic, provided the Idea is good." Just like spreading democracy or putting a man on the moon, breaking oil's monopoly in the transportation sector and the creation of an energy transportation revolution by turning oil into salt is a big idea, one that could greatly improve the human condition, our prosperity, and our national security. It requires dedicated and enthusiastic leadership, sustained public support and close international cooperation, and, of course, money. But perhaps more than anything it requires that we as a nation decide that we want Churchill's words to be those that depict us in the annals of history, not Juvenal's. The choice to have choice is ours and ours alone.

⌘ ⌘ ⌘

INTRODUCTION:
FROM THE BEACH TO THE BASEBALL DIAMOND

[1] "Energy: The Albatross of National Security," Leadership Forum address by Senator Richard Lugar at the Brookings Institution, March 13, 2006,
http://lugar.senate.gov/energy/press/articles/060301cepquarterly.cfm

[2] "Crude Awakening: Saudi Oil Minister Warns Against Renewable Exuberance," *Wall Street Journal*, February 11, 2009.

[3] "Energy Agency Sets Grim Oil Forecast," *Wall Street Journal*, November 8, 2005.

[4] *The Energy Learning Curve*, Report from Public Agenda, April 2009.
http://www.publicagenda.org/files/pdf/energy_learning_curve.pdf

[5] *World Energy Outlook 2008*, International Energy Agency, 2008.

CHAPTER 1
UNDERSTANDING STRATEGIC COMMODITIES

[1] Mark Kurlansky, *Salt: A World History*, (NY: Penguin, 2002), 6.

[2] Ibid, 258.

[3] Ibid, 35.

[4] Ibid, 141.

[5] Ibid, 207.

[6] Ibid, 222.

[7] Ibid, 12-13.

[8] Edwin Black, *Internal Combustion: How Corporations and Governments Addicted the World to Oil and Derailed the Alternatives*, (NY: St. Martin's Press, 2006)

[9] Barbara Freese, *Coal, A Human History*, (NY: Perseus, 2003)

[10] Daniel Yergin, *The Prize: The Epic Quest for Oil, Money and Power*, (NY: Touchstone 1993), 544

[11] "As China, U.S. Vie for More Oil, Diplomatic Friction May Follow," *Washington Post*, April 16, 2006.

CHAPTER 2
WHAT IS ENERGY INDEPENDENCE ANYWAY?

[1] National Petroleum Council (NPC), *Facing the Hard Truths About Energy*, (Washington. D.C.: NPC, July 18, 2007), http://www.npchardtruthsreport.org.

[2] Robert Bryce, *Gusher of Lies: The Dangerous Delusions of Energy Independence*, (Jackson, TN: Public Affairs, 2008)

[3] Daniel Yergin, "Ensuring Energy Security," *Foreign Affairs*, March/April 2006.

[4] James R. Schlesinger, "Thinking Seriously About Energy and Oil's Future," *The National Interest*, Winter 2005/06.

[5] Roger Sant and Michael Kinsley, "Why Energy Independence?," *Washington Post*, December 14, 2008.

[6] Irwin Stelzer, *Energy Policy: Abandon Hope All Ye Who Enter Here*, Hudson Institute, Summer 2008, http://www.hudson.org/files/publications/Stelzer.pdf

[7] Jerry Taylor, "Energy Independence? Kerry's Dreaming," Cato Institute, August 24, 2004, http://www.cato.org/pub_display.php?pub_id=2796

[8] "Frank Verrastro and Sarah Ladislaw, "Providing Energy Security in an Interdependent World," *The Washington Quarterly*, Autumn 2007, http://www.twq.com/07autumn/docs/07autumn_verrastro.pdf

[9] Robert Manning, "Energy Independence Fallacy," *New Atlanticist*, January 5, 2009, http://www.acus.org/new_atlanticist/energy-independence-fallacy

[10] *US Energy Policy FAQ: The U.S. Energy Mix, National Security, and the Myths of Energy Independence*, The James A. Baker Institute for Public Policy, Rice University, February 1, 2008.

[11] Flynt Leverett, comments in The Atlantic Magazine Green Intelligence Forum, November 19, 2008.

[12] Remarks by Charles Wald at a panel on "Energy, the Environment and National Security" at the Center for American Progress/Century Foundation Conference, America in the World, Washington DC, June 12, 2007.

[13] *National Security Consequences of U.S. Oil Dependence*, Council on Foreign Relations Task Force, 2006, 4. http://www.cfr.org/content/publications/attachments/EnergyTFR.pdf

[14] Verrastro and Ladislaw.

[15] Sant and Kinsley.

[16] Andy Grove, "Our Electric Future," *The American Magazine*, July/August 2008.

[17] "Flapping Their Gum: Sudan Threatens U.S. Soft Drinks," *ABC News*, June 1, 2007, http://abcnews.go.com/US/story?id=3232434

[18] Eric Lichtblau, "Documents Back Saudi Link to Extremism," *New York Times*, June 23, 2009. http://www.nytimes.com/2009/06/24/world/middleeast/24saudi.html?em

[19] "Who Are the Foreign Fighters in Iraq?," *NBC News*, June 20, 2005.

[20] "New look at Foreign Fighters in Iraq," *Christian Science Monitor*, January 7, 2008.

[21] According to a 1998 sworn statement of an Afghani witness, an emissary for Prince Turki al-Faisal handed a check for one billion Saudi riyals to a top Taliban leader. Eric Lichtblau, "Documents Back Saudi Link to Extremism," *New York Times*, June 23, 2009. http://www.nytimes.com/2009/06/24/world/middleeast/24saudi.html?em

[22] "Obama to Face 'More Radicalised' Middle East," *The Time*, January 16, 2009.

[23] Prince Turki al-Faisal, "Saudi Patience is Running Out," *Financial Times*, January 22, 2009.

[24] Thomas Friedman, "Drowning Freedom in Oil," *New York Times*, August 25, 2002.

[25] "Libya Halts Swiss Oil Shipments," *BBC*, July 24, 2008, http://news.bbc.co.uk/2/hi/europe/7523537.stm

CHAPTER 3
PLANS APLENTY, MOSTLY EMPTY

[1] *Report of the National Energy Policy Development Group*, White House, http://www. whitehouse.gov/energy/2001/index.html, and *National Security Consequences of U.S. Oil Dependence*, Council on Foreign Relations, Independent Task Force Report No. 58 http://www.cfr.org/content/publications/attachments/EnergyTFR.pdf

[2] Thomas L. Friedman, "Making America Stupid," *New York Times*, September 13, 2008.

[3] "Analysis of Crude Oil Production in the Arctic National Wildlife Refuge," EIA Report #: SR-OIAF/2008-03, May 2008, http://www.eia.doe.gov/oiaf/servicerpt/anwr/results. html

[4] "Saudis Put Oil Capacity Rise on Hold," *Financial Times*, April 21, 2008

[5] Jim Kliesch, *Setting the Standard: How Cost-Effective Technology Can Increase Vehicle Fuel Economy*, Union of Concerned Scientists, September 2008.

[6] Robert Zubrin, *Energy Victory,* 24.

[7] Thomas Friedman, "Win, Win, Win, Win," *New York Times*, December 27, 2008.

[8] Charles Krauthammer, The Net Zero Gas Tax, *Weekly Standard*, January 5, 2009, http://www.weeklystandard.com/Content/Public/Articles/000/000/015/949rsrgi.asp

[9] Ibid.

[10] "The Soviet effort to dominate Afghanistan has brought Soviet military forces to within 300 miles of the Indian Ocean and close to the Straits of Hormuz, a waterway through which most of the world's oil must flow. The Soviet Union is now attempting to consolidate a strategic position, therefore, that poses a grave threat to the free movement of Middle East oil. [...] Let our position be absolutely clear: An attempt by any outside force to gain control of the Persian Gulf region will be regarded as an assault on the vital interests of the United States of America, and such an assault will be repelled by any means necessary, including military force." President Jimmy Carter, State of the Union Address, January 23, 1980

[11] Vijay Vaitheeswaran, *Power to the People: How the Coming Energy Revolution Will Transform an Industry, Change Our Lives, and Maybe Even Save the Planet*, (NY: Farrar, Straus and Giroux, 2003), 260.

[12] "Administration Shifts Strategy on Auto Fuels," *New York Times*, January 9, 2002.

[13] "Next Energy: Powering Michigan Future," Speech by Governor Engler, Dearborn Michigan, April 18, 2002, http://michigan.michigan.gov/formergovernors/0,1607,7-212-31303_31317-33550--,00.html

[14] Jeremy Rifkin, *The Hydrogen Economy: The Next Great Economic Revolution*, (NY: Tarcher/ Putnam 2002).

[15] Matthew Wald, "U.S. Drops Research Into Fuel Cells for Cars," *New York Times*, May 7, 2009.

CHAPTER 4
FROM THE CAR AHMADINEJAD LOVES TO
THE ONE BIN LADEN HATES

[1] Gal Luft and Anne Korin, *Ahmadinejad's Gas Revolution: A Plan to Defeat Economic Sanctions*, IAGS report, December 2006, http://www.iags.org/iran121206.pdf

[2] Natural Gas Vehicle Coalition Newsletter, March 6, 2009, Volume 12, Edition 9.

[3] Greg Dolan, "China Takes Gold in Methanol Fuel," *Journal of Energy Security*, October 2008.

[4] "Chu: All New Cars Should have Flex-fuel Capacity," *De Moines Registrar*, June 22, 2009.

[5] *Pocket Guide to Transportation 2008*, Bureau of Transportation Statistics, U.S. Department of Transportation.

[6] "Thousands Fleeing Rita Jam Roads from Coast," *Washington Post*, September 23, 2005.

[7] Notice we didn't write 100-150 miles per gasoline equivalent gallon – we aren't converting the grid electricity used to power the plug in to a gasoline equivalent. Instead, we're assessing how far electricity stretches each gallon of gasoline. To avoid confusion we use the abbreviation mpgg rather than mpg.

[8] Sherry Boschert, *Plug-in Hybrids: The Cars that will Recharge America*, (Gabriola Island, BC: New Society Publishers, 2006).

CHAPTER 5
FUELING 48 FLOORS

[1] The number following the B indicates the proportion of biodiesel in a biodiesel and petroleum diesel blend. So B5 is a blend with 5% biodiesel and 95% petroleum diesel. B100 is pure biodiesel.

[2] New Biodiesel Blend Specifications Published by ASTM International, National Biodiesel Board Press Release, October 14, 2008.

[3] Municipal Solid Waste in the U.S. 2007 Fact and Figures, U.S. Environmental Protection Agency, http://www.epa.gov/osw/nonhaz/municipal/pubs/msw07-rpt.pdf

[4] Brad Lemley, "Anything into Oil," *Discover Magazine*, April 2006, http://discovermagazine.com/2006/apr/anything-oil

[5] Pocket Guide to Transportation 2008, Bureau of Transportation Statistics

[6] U.S. Census Bureau, 2005-2007 American Community Survey, S0802. Means of Transportation to Work by Selected Characteristics, Data Set: 2005-2007

[7] *Impacts Assessment of Plug-in Hybrid Vehicles on Electric Utilities and Regional Power Grids*, Pacific Northwest National Laboratory, November, 2007.

[8] *Potential Impacts of Plug-in Hybrid Electric Vehicles on Regional Power Generation*, Oak Ridge National Laboratory January 2008. http://www.ornl.gov/info/ornlreview/v41_1_08/regional_phev_analysis.pdf

[9] Biomass Resource Estimates, Oak Ridge National Laboratory, http://bioenergy.ornl.gov/papers/misc/resource_estimates.html

[10] *Biomass as a Feedstock for Bioenergy and Bioproducts Industry: The Technical Feasibility of an Annual Billion Ton Annual Supply*, (Washington, DC: U.S. Department of Energy, April 2005), http://feedstockreview.ornl.gov/pdf/billion_ton_vision.pdf

[11] Zhu Xifeng, Biomass Pyrolysis and its Potential for China, International Conference on Bioenergy Utilization and Environment Protection –6th LAMNET Project Workshop, 24 - 26 September 2003, Dalian, China.

[12] "The 5% Solution: Methanol Production Feedstock Diversity", Methanol Institute factsheet

[13] Kevin Hassett, "Ethanol's a Big Scam, and Bush Has Fallen for It," *Bloomberg*, February 13, 2006.

[14] "Everyone Hates Ethanol," *Wall Street Journal*, March 16, 2009.

[15] Michael Grunwald, "The Clean Energy Scam," *Time Magazine*, March 27, 2008.

[16] Lester Brown, "Biofuels Blunder, Massive Diversion of U.S. Grain to Fuel Cars is Raising World Food Prices, Risking Political Instability," Briefing before U.S. Senate Committee on Environment and Public Works, June 13, 2007, http://www.earth-policy.org/Transcripts/SenateEPW07.htm

[17] "Biofuels Crime against Humanity," *BBC News*, October 27, 2007, http://news.bbc.co.uk/2/hi/americas/7065061.stm

[18] "Venezuela's Chavez Slams Bush Ethanol Plan," *Reuters*, April 10, 2007.

[19] "Saudi scholar warns alcohol in bio fuel is a sin," *AlArabiya.net*, February 19, 2009, http://www.alarabiya.net/articles/2009/02/19/66803.html

[20] "OPEC President Blames Ethanol from Crude Price Rise," *Marketwatch*, July 6, 2008, http://www.marketwatch.com/story/opec-president-blames-oil-prices-on-ethanol-weak-dollar-reports

[21] Doug Koplow, *Subsidies in the US Energy Sector: Magnitude, Causes, and Options for Reform*, Earth Track, (Cambridge, MA), November 2006, http://www.earthtrack.net/earthtrack/library/SubsidyReformOptions.pdf

[22] C. Boyden Gray and Andrew R. Varcoe, "Octane, Clean Air, and Renewable Fuels: A Modest Step toward Energy Independence," *Texas Review of Law & Politics, Vol.10, 2006.* http://www.trolp.org/main_pgs/issues/v10n1/Gray.pdf

[23] "As Biofuels Catch On, Next Task Is to Deal With Environmental, Economic Impact," *Wall Street Journal*, March 24, 2008.

[24] Michael Pollan, "Farmer in Chief," *New York Times Magazine*, October 9, 2008, http://www.nytimes.com/2008/10/12/magazine/12policy-t.html

[25] Bruce Dale, "Biofuels: Thinking Clearly about the Issues," in Daveed Gartenstein Ross and Clifford May eds. *From Energy Crisis to Energy Security*, (Washington, DC: FDD Press, 2008), p.57.

[26] See also Hosein Shapouri, James A. Duffield, and Michael Wang, "The Energy Balance of Corn Ethanol: An Update," U.S. Department of Agriculture, Office of the Chief Economist, Office of Energy Policy and New Uses, Agricultural Economic Report No. 814, http://www.usda.gov/oce/reports/energy/aer-814.pdf

[27] Tim Searchinger et al, "Use of U.S. croplands for biofuels increases greenhouse gases through emissions from land-use change," *Science Magazine*, February 29, 2008, p. 1238.

[28] Robert Zubrin, "The Irrationality of Indirect Analysis," *Roll Call*, June 3, 2009, http://www.rollcall.com/news/35481-1.html

[29] Michael Wang, The Debate on Energy and Greenhouse Gas Emissions Impacts of Fuel Ethanol, Argonne National Laboratory, August 3, 2005, http://www.transportation.anl.gov/pdfs/TA/347.pdf

[30] *An Examination of the Potential for Improving Carbon/Energy Balance of Bioethanol*, A Report to IEA Bioenergy Task Force 39, February 15, 2009

[31] The Impact of Ethanol Use on Food Prices and Greenhouse Gas Emissions, Congressional Budget Office, April 2009.

[32] http://www.solixbiofuels.com/html/why_algae.html

[33] Biofuels study sees 90 billion gallons by 2030, *Associated Press*, February 10, 2009

[34] Senator Charles Grassley, Lifting Ethanol Tariff won't Lower Gas Prices for Consumers, Press Release, May 4, 2006. http://grassley.senate.gov/news/Article.cfm?customel_dataPageID_1502=10847

CHAPTER 6
MELTING ICE MEETS MELTING WEST: POWERING OUR CARS WITH CO_2

[1] "Kids Fear Global Warming More than Terrorism, Car Crashes and Cancer, According to a National Earth Day Survey," National Survey of Middle School Students, April 20, 2007.

[2] Dan Vergano, "The Debate's Over: Globe *is* Warming," *USA Today*, June 13, 2005.

[3] Time Magazine Special Report Global Warming, April 3, 2006.

[4] Lawrence Wright, *The Looming Tower: Al Qaeda and the Road to 9/11,* (NY: Knopf, 2006), 149.

[5] Thomas L. Friedman, *Hot Flat and Crowded: Why We Need a Green Revolution and How It Can Renew America*, (NY: Macmillan, 2008), 82.

[6] *The Future of Coal*, MIT Report, http://web.mit.edu/coal/

[7] "Japan's Mitsui Chemicals to Make Methanol from CO_2," *Reuters*, August 28, 2008

[8] Mark Allen Bernstein, *A Viable Option for Biofuel Feedstock*, May 1, 2007, a paper commissioned by GreenFuel Technologies Corporation and was researched and written under the auspices of the University of Southern California.

CHAPTER 7
THINKING *INSIDE* THE BOX: WHY BATTERIES ARE SO EXPENSIVE AND WHAT WE CAN DO TO BRING DOWN THEIR COST

[1] Ken Thomas, "Obama Plug-in Cars Goal Hard to Hit in the U.S.," *U.S. News and World Report*, April 15, 2009, http://www.usnews.com/articles/science/2009/04/15/obamas-plug-in-cars-goal-hard-to-hit-in-us.html

[2] President George Bush's speech at the Dupont Experimental Station in Delaware, January 24, 2007.

[3] "Israel Looks to Electric Cars," *Time Magazine*, January 20, 2008.

[4] "Hawaii Makes Big Bet on Electric Cars," *Wall Street Journal*, December 2, 2008.

[5] "Google Plugs In to Hybrid Car Development with $10M," *USA Today*, June 22, 2007.

[6] Dan Reicher, "Commentary," in David Sandalow ed. *Plug-in Electric Vehicles: What Role for Washington?* (Washington, DC: Brookings Press, 2009), xiii

[7] Clean Energy 2030, Google's Proposal for Reducing U.S. Dependence on Fossil Fuels, http://knol.google.com/k/-/-/15x31uzlqeo5n/1#

[8] Cited in *Internal Combustion*.

[9] U.S. Department of Homeland Security, U.S. Fire Administration, Highway Vehicle Fires, http://www.usfa.dhs.gov/downloads/pdf/tfrs/v9i1.pdf

[10] "Guess Who Hopes to Help Power New Hybrid Cars," *Wall Street Journal*, March 11, 2008.

[11] "McCain calls for $300 million prize for better car battery," *CNN*, June 24, 2008

[12] Ted J. Miller, Automotive Requirements for Lithium Batteries, Ford Motor Company, January 27, 2009.

[13] "GM's Volt to Debut in Washington, Bay Area," *Washington Post*, February 5, 2009.

[14] "Volkswagen Eyes China Venture," *Wall Street Journal*, May 27, 2009
http://online.wsj.com/article/SB124331239762553635.html

[15] A Congressional authorization provides permission for a government expenditure, but doesn't actually allocate funds for that purpose. In order to allocate the funds, Congress needs to pass an appropriation. Congress passes lots of authorizations for which Members of Congress take credit for "doing something," but unless Congress actually appropriates the funds, authorizations are meaningless.

[16] The best summary of the legislative action on electric transportation can be found in Dean Taylor, "Current Federal Authorized Programs on Plug-In Hybrids, Battery Electric Vehicles, and Related Efforts," in David Sandalow ed. *Plug-in Electric Vehicles: What Role for Washington?* (Washington, DC: Brookings Press, 2009),

[17] Charles W. Wessner, *Securing the Future: Regional and National Programs to Support the Semiconductor Industry*, (National Academies Press, 2003)

[18] Jack Lifton, "The Future of Nickel Metal Hydride Battery and the Rare Earth Metals it is Constructed From," *Resource investor*, January 23, 2009, http://www.resourceinvestor.com/pebble.asp?relid=48977

[19] "A Blended Battery Pack for Cars Combining Different Battery Technologies Could Improve Vehicle Performance and Reduce Costs," *Technology Review*, January 26, 2009. http://www.technologyreview.com/energy/22015/

CHAPTER 8
BACK TO THE SALT BRINES: TOMMOROW'S DEPENDENCIES

[1] Brendan I. Koerner, "The Saudi Arabia of Lithium," *Forbes*, November 24, 2008.

[2] Bjorn Andersson and Inge Rade, "Metal Resource Constraints for Electric Vehicle Batteries," Transportation Research Part D 6, 297-324, 2001.

[3] http://lithiumabundance.blogspot.com

[4] "Bolivia Holds Key to Electric Car Future," *BBC News*, November 9, 2008, http://news.bbc.co.uk/2/hi/business/7707847.stm

[5] "Bolivia Pins Hopes on Lithium, Electric Vehicles," *Associated Press*, March 1, 2009, http://www.msnbc.msn.com/id/29445248/

[6] USGS, http://minerals.usgs.gov/minerals/pubs/commodity/lithium/mcs-2008-lithi.pdf

[7] Irving Mintzer, "Look Before You Leap: Exploring the Implications of Advanced Vehicles for Import Dependence and Passenger Safety," in David Sandalow ed. *Plug-in Electric Vehicles: What Role for Washington?* (Washington, DC: Brookings Press, 2009), 111.

[8] Ibid, 116.

[9] Ibid, 120.

[10] "Lynas Corporation Benefits from Chinese Rare Earth Export Reduction, Proactive Investors," http://www.proactiveinvestors.com.au/companies/news/626/lynas-corporation-benefits-from-chinese-rare-earth-export-reduction-0626.html

[11] U.S. Geological Survey, http://minerals.usgs.gov/minerals/pubs/commodity/rare_earths/mcs-2008-raree.pdf

⌘　⌘　⌘

INDEX

Made in the USA
Lexington, KY
08 June 2010